A LETTER FROM JACK DANGERMOND

Given the rapid pace of human impact on their surroundings, it remains clear that our world is increasingly environmentally challenged, and humans must take more responsibility for creating a sustainable future. This collection of maps illustrates the interesting work of the GIS community. They tell stories and show how organizations around the world address diverse and complex geographic problems. Although humans have been around a long time, our world is still not well known. Many of these maps reveal previously unknown information and, in some cases, identify a problem and provide greater context. Even more interesting are the maps furnishing a decision-making framework.

One of these efforts, featured on the cover, is a visualization by the Half-Earth Project, a program of the E. O. Wilson Foundation. The layer shown indicates the scale of land conservation needed to protect Earth's biodiversity. Despite E. O. Wilson's passing, his call continues for us to truly consider the value of our planet and preserve it for future generations.

Collectively, we as humans, and particularly as GIS users, possess the tools to consider the future impact of our work. Our capacity to make better choices is unprecedented. Taken as a whole, *Esri Map Book*, *Volume 37*, offers a vision of the world as a single ecosystem. Consider the insights that would be possible if all of these maps, and millions like them, along with their underlying data and the unique perspectives of their authors were brought together and used to address the great sustainability challenges of our time. This is happening rapidly and, to my mind, will be essential for our future. I refer to this as *geospatial infrastructure*, a new pattern of GIS deployment where users share their work and allow others to easily discover and integrate social, economic, and environmental data from many sources. The vision of a global geospatial platform is emerging rapidly on the web, and is already being used by millions of professionals as a *living atlas of our planet*.

Organizations are increasing applying GIS and the *geographic approach* as a holistic way of thinking and collaborative problem solving. In the future, this technology and approach will need to be integrated into every sector of society, facilitating an inclusive and more sustainable future.

Warm regards,

Jack Dangermond

CONTENTS

HAIL TRENDS IN THE UNITED STATES

Property & Liability Resource Bureau
Downers Grove, Illinois, USA
By Andrew Louchios

Historical hail reports can help insurance companies identify areas that are at a higher risk of hail damage. Reports of 1-inch hail or larger from 1995 through 2019 were used to depict hail trends in the contiguous United States. There is typically minimal to no damage to roofing materials and vehicles from smaller hail.

A notable trend is seen in much of the Plains with intensifying hot spots. These indicate a statistically significant hot spot in which the intensity of clustering of high counts of hail reports in each period is increasing. The hail reports in this region could signify climate change or increased reporting because of population growth combined with easier ways to report hail due to technological advances (widespread internet use, social media, and apps such as mPING).

CONTACT
Andrew Louchios
alouchios@plrb.org

SOFTWARE
ArcGIS® Desktop

DATA SOURCES
National Weather Service Storm Prediction Center

Courtesy of Property & Liability Resource Bureau.

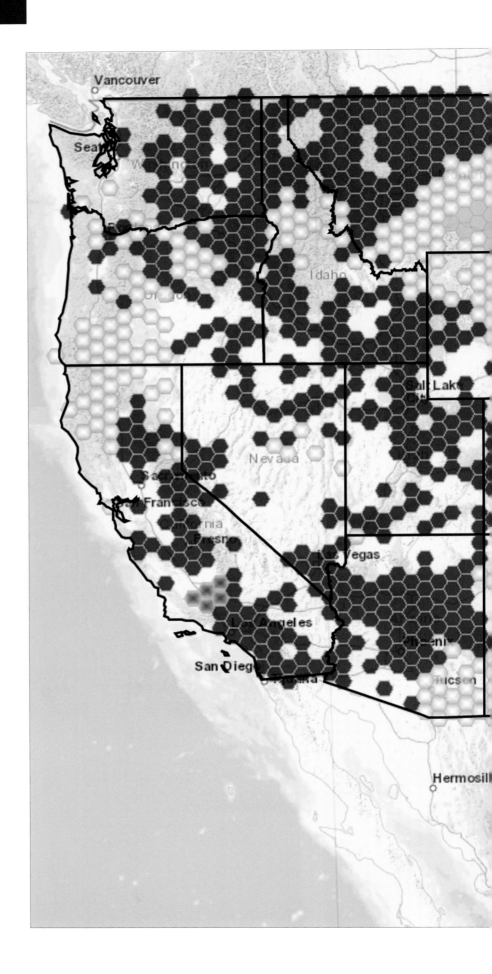

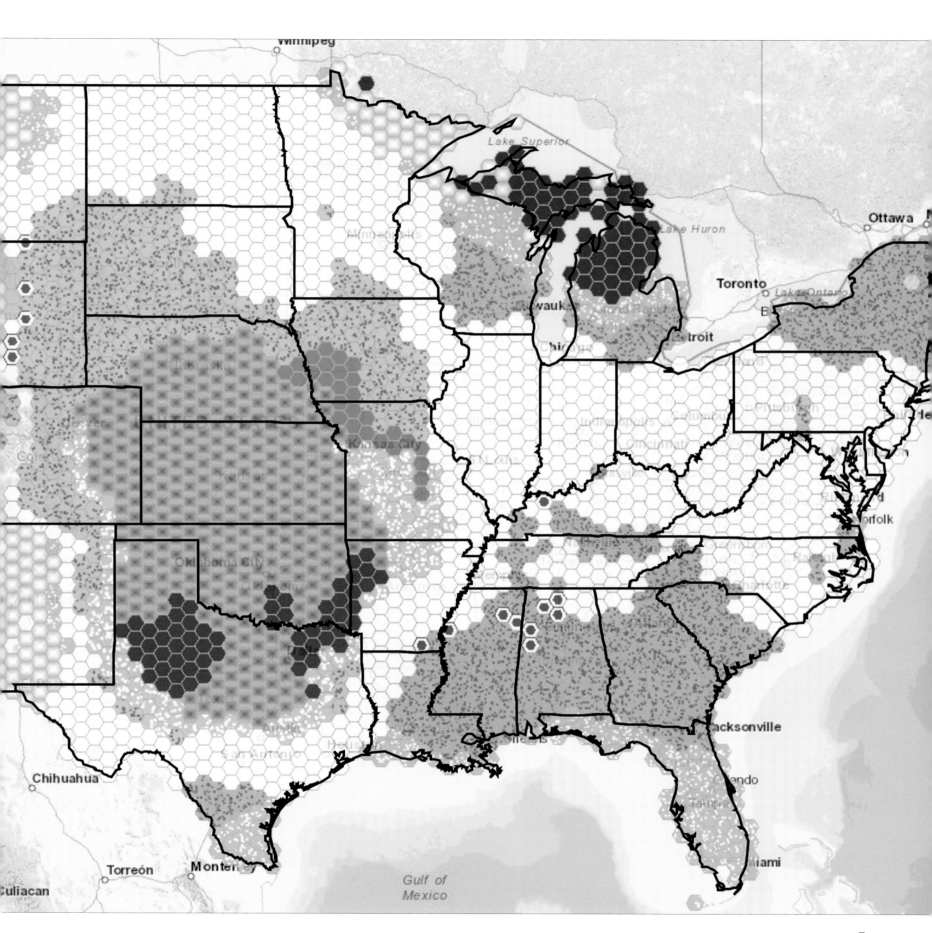

NATURAL DISASTER RISKS IN JAPAN

Tokio Marine dR Co., Ltd., Tokyo, Japan
By Masako Ikeda

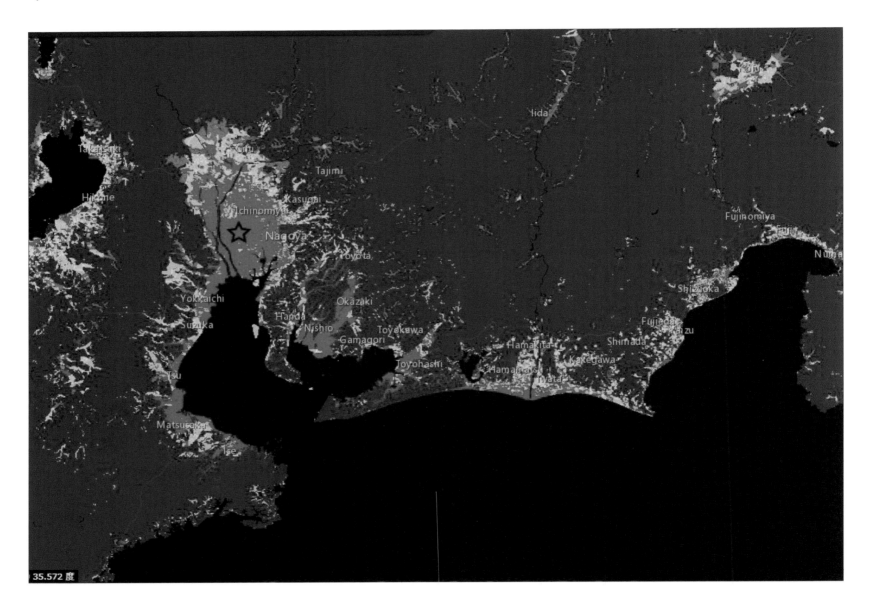

These maps show the Analytical Location Intelligence Service (ALIS), an application that centralizes data related to natural disasters and visualizes it on a map. The maps indicate estimated damage due to liquefaction in the event of a Nankai Trough Mega Earthquake and areas susceptible to flooding due to a heavy rainfall event.

Tokio Marine dR Co., Ltd. (TdR) provides consulting services for various risks affecting companies, such as natural disasters and management issues. This one-stop system for searching, extracting, and delivering information on multiple potential natural disasters at a client's property has not only improved the efficiency and quality of work, but has also dramatically increased the number of internally generated reports that include hazard maps.

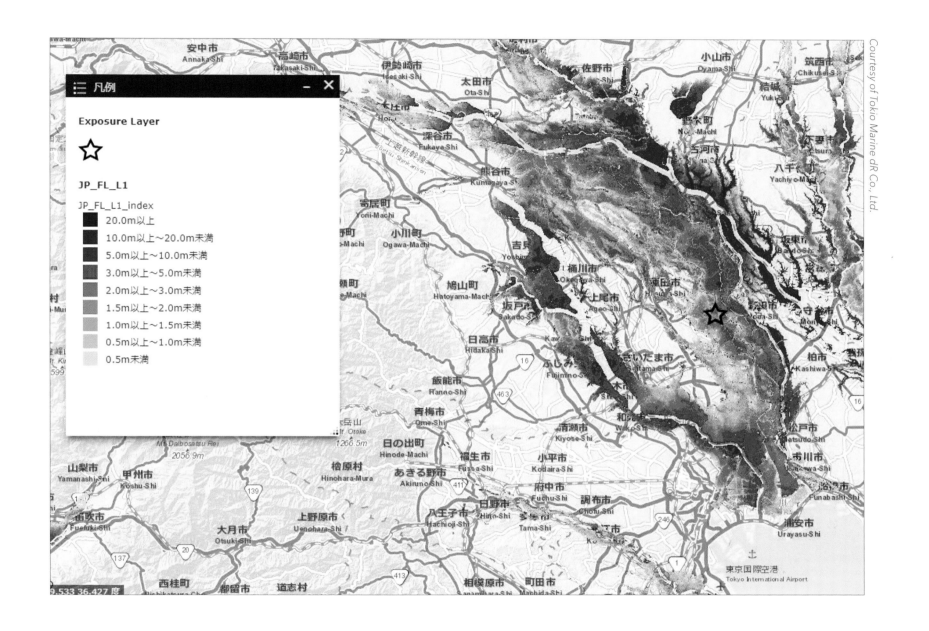

CONTACT
Tokio Marine dR Co., Ltd.
contact.ddg@tokiorisk.co.jp

SOFTWARE
ArcGIS Pro, ArcGIS Online,
ArcGIS Enterprise

DATA SOURCES
Map of possible flood zone (prepared by TdR based on data provided
by the Ministry of Land, Infrastructure, Transport, and Tourism)

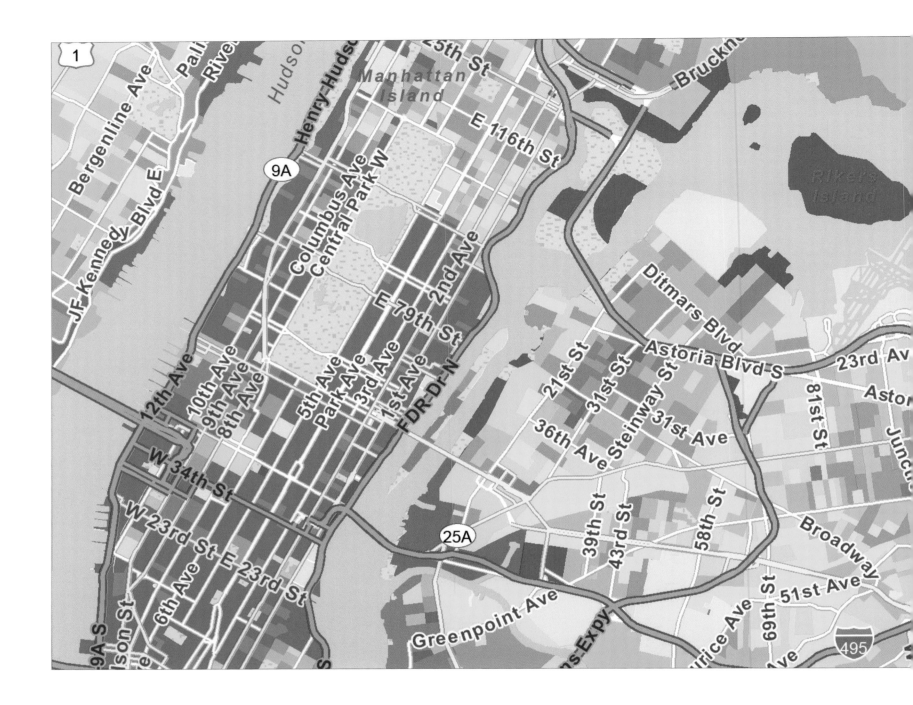

WEATHERING INCOME SHOCK

Retail Profit Management (RPM)
Santa Clarita, California, USA
By Steve Lackow

How well prepared are you and your family to withstand a sudden loss of income? The Federal Reserve Board asked this question of Americans in the latest Survey of Consumer Finances (SCF). Using RPM's proprietary MarketBank technology and market populations estimated by Esri®, RPM projected the SCF results to census block groups, and the result is this map of resiliency, showing areas where the population is prepared to weather income shock, and those areas where people are less prepared and might need assistance.

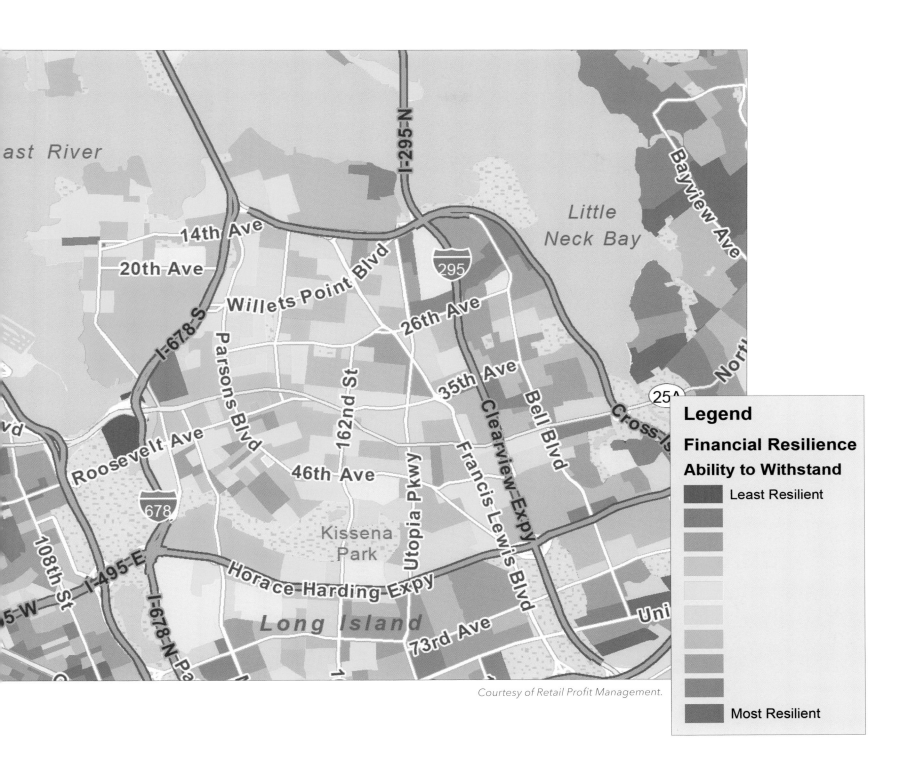

Courtesy of Retail Profit Management.

Legend

Financial Resilience

Ability to Withstand

Least Resilient

Most Resilient

CONTACT
Steve Lackow
slackow@rpmconsulting.com

SOFTWARE
ArcGIS Desktop,
ArcGIS® Business Analyst™ Desktop

DATA SOURCES
RPM MarketBank, ArcGIS Business Analyst,
Federal Reserve Bank Survey of Consumer Finances

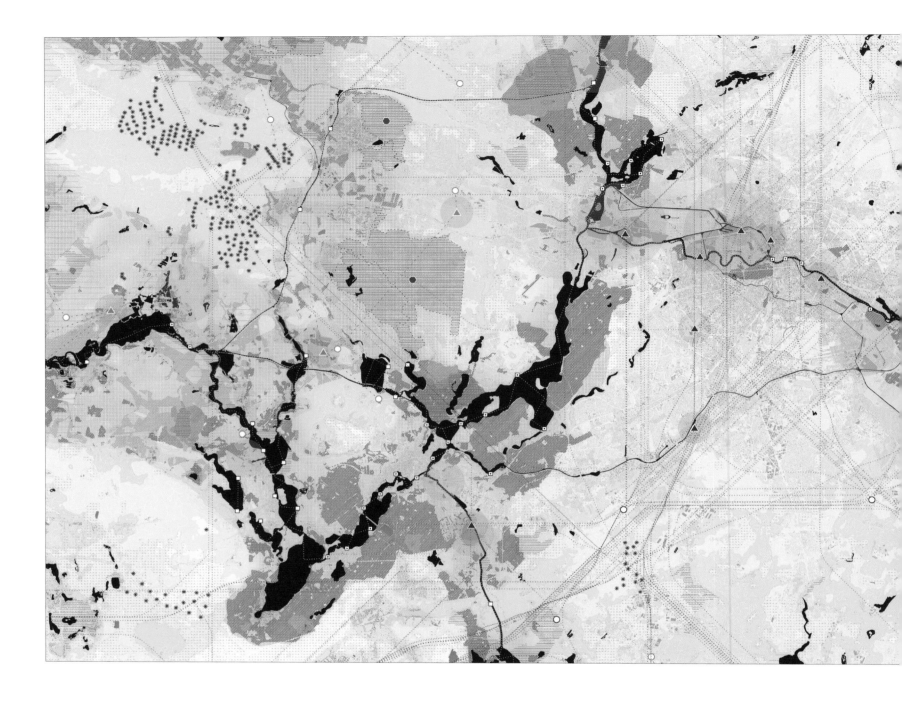

BERLIN-BRANDENBURG 2070 REGIONAL DEVELOPMENT FRAMEWORK

TSPA
Berlin-Brandenburg,
Germany
By Filippo Imberti

This map shows the territorial transformation of the Berlin-Brandenburg region. It offers a rethinking of the existing structures and relations between cities and nature. This remapping transforms the traditional rural to urban linear flow of energy and natural resources into circular processes for ecosystem services and renewable energy production, ensuring a resilient and productive future for the German capital and its broader territory.

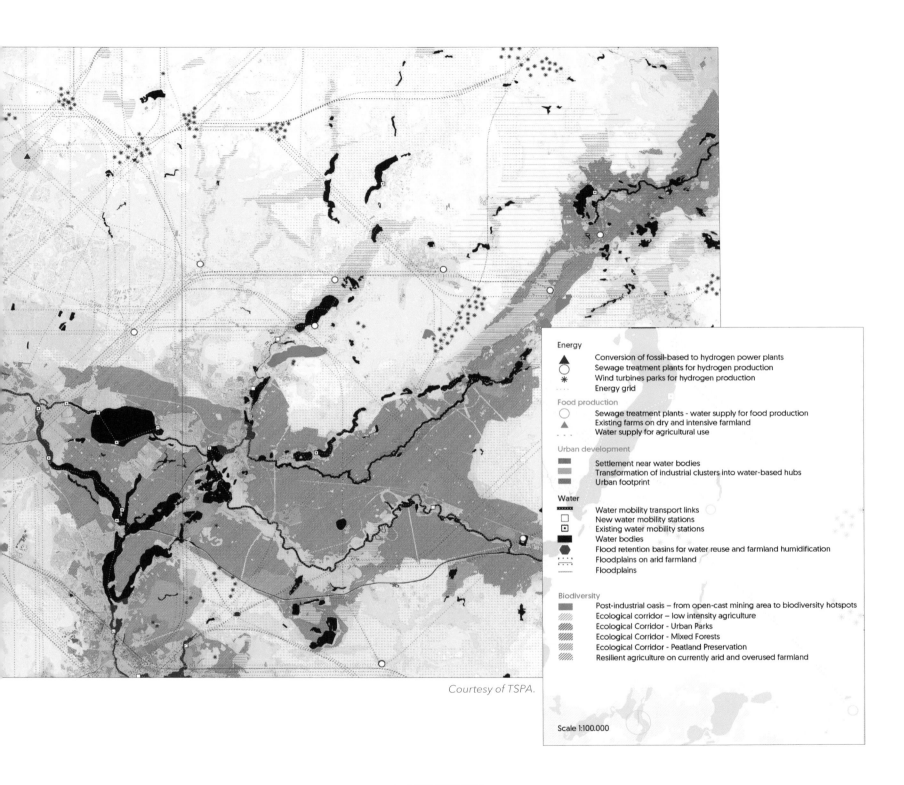

Energy

▲ Conversion of fossil-based to hydrogen power plants
○ Sewage treatment plants for hydrogen production
✳ Wind turbines parks for hydrogen production
···· Energy grid

Food production

○ Sewage treatment plants - water supply for food production
▲ Existing farms on dry and intensive farmland
-·-· Water supply for agricultural use

Urban development

Settlement near water bodies
Transformation of industrial clusters into water-based hubs
Urban footprint

Water

Water mobility transport links ○
□ New water mobility stations
▣ Existing water mobility stations
█ Water bodies
⬡ Flood retention basins for water reuse and farmland humidification
∴ Floodplains on arid farmland
Floodplains

Biodiversity

Post-industrial oasis – from open-cast mining area to biodiversity hotspots
Ecological corridor – low intensity agriculture
Ecological Corridor - Urban Parks
Ecological Corridor - Mixed Forests
Ecological Corridor - Peatland Preservation
Resilient agriculture on currently arid and overused farmland

Courtesy of TSPA.

Scale 1:100.000

CONTACT
Filippo Imberti
fi@tspa.eu

SOFTWARE
ArcGIS Desktop, ArcGIS Pro,
ArcGIS Living Atlas of the World,
ArcGIS® Spatial Analyst™

DATA SOURCES
Landesvermessung und Geobasisinformation Brandenburg (LGB),
OpenStreetMap

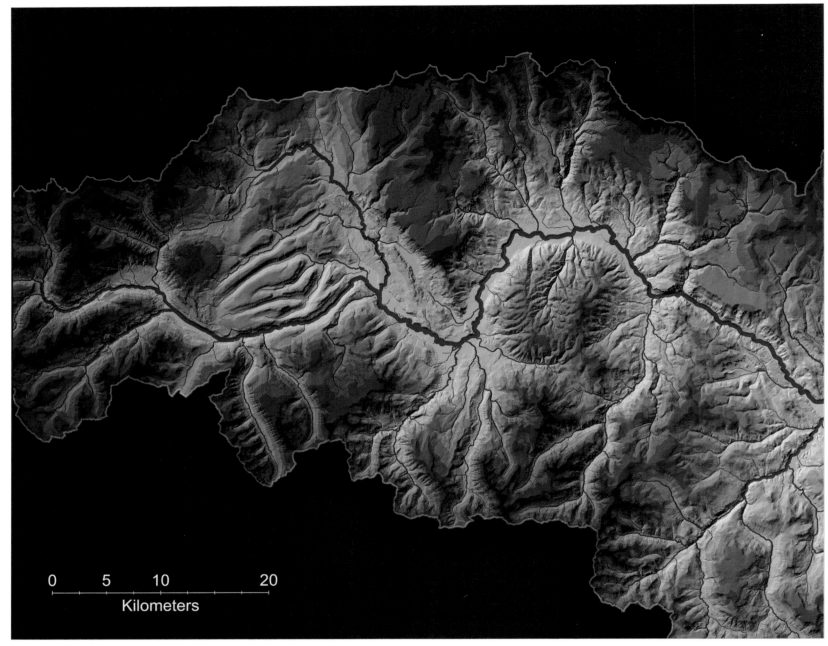

Courtesy of SterlingCarto LLC.

HYDROLOGY OF THE RIO GRANDE HEADWATERS

SterlingCarto LLC and AEGIS
Technologies
Westminster, Colorado, USA
By Sterling Loetz and Christina Kumar

The Rio Grande headwater catchment basin collects and drains to one of the longest river systems in North America. Located in the San Juan Mountain range in southern Colorado, the basin expands over an area of approximately 1,026,120 acres (4,153 sq km).

This map outlines the catchment basin using elevation-derived modeling to show how the topography guides runoff toward the eastern edge of the basin from high to low elevations.

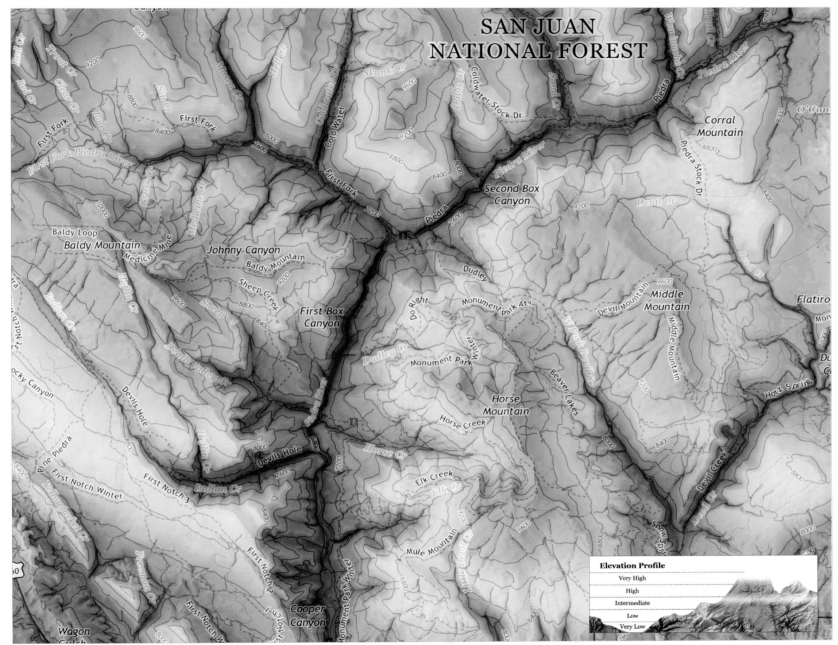

Elevation Profile

Very High
High
Intermediate
Low
Very Low

Courtesy of SterlingCarto LLC .

US Geological Survey (USGS)

ELEVATION MAP OF THE WEMINUCHE WILDERNESS, COLORADO

This map displays topography and shaded terrain across the Weminuche Wilderness and San Juan National Forest in southwest Colorado. A combination of digital terrain data and hillshade processing techniques are used to accentuate the bare earth terrain across the landscape.

Prominent visual characteristics such as mountain peaks, valleys, plateaus, plains, and fault lines, all produced from geologic processes, can be discerned from the elevation data alone.

CONTACT
Sterling Loetz
sterlingcarto@gmail.com

SOFTWARE
ArcGIS Pro

DATA SOURCES
US Geological Survey (USGS)

SALZKAMMERGUT ATLAS

Freytag & Berndt and Artaria KG
Vienna, Austria
By Hannes Mittergeber, Albin Tempelmayr, Bernhard Brabenec, Peter Lipták, and Bettina Wobek

Freytag & Berndt is Austria's biggest cartographic publisher. For the company's 250th anniversary, a new series of Austrian hiking atlases was launched. The hiking atlas of the Salzkammergut region has a scale of 1:40,000 and is 240 pages long. It highlights 40 trails across the region with detailed maps supplemented by altitude profiles, brief information, and map excerpts. The book includes hiking tours of the very popular Austrian lakes of the region and many beautiful peaks such as the Krippenstein, Schafberg, Traunstein, Loser and many more with amazing views.

CONTACT
Hannes Mittergeber
hannes.mittergeber@freytagberndt.com

SOFTWARE
ArcGIS Pro

DATA SOURCES
Internal, DEM 10m Austria (CC BY 3.0 AT)

Courtesy of Freytag & Berndt and Artaria KG.

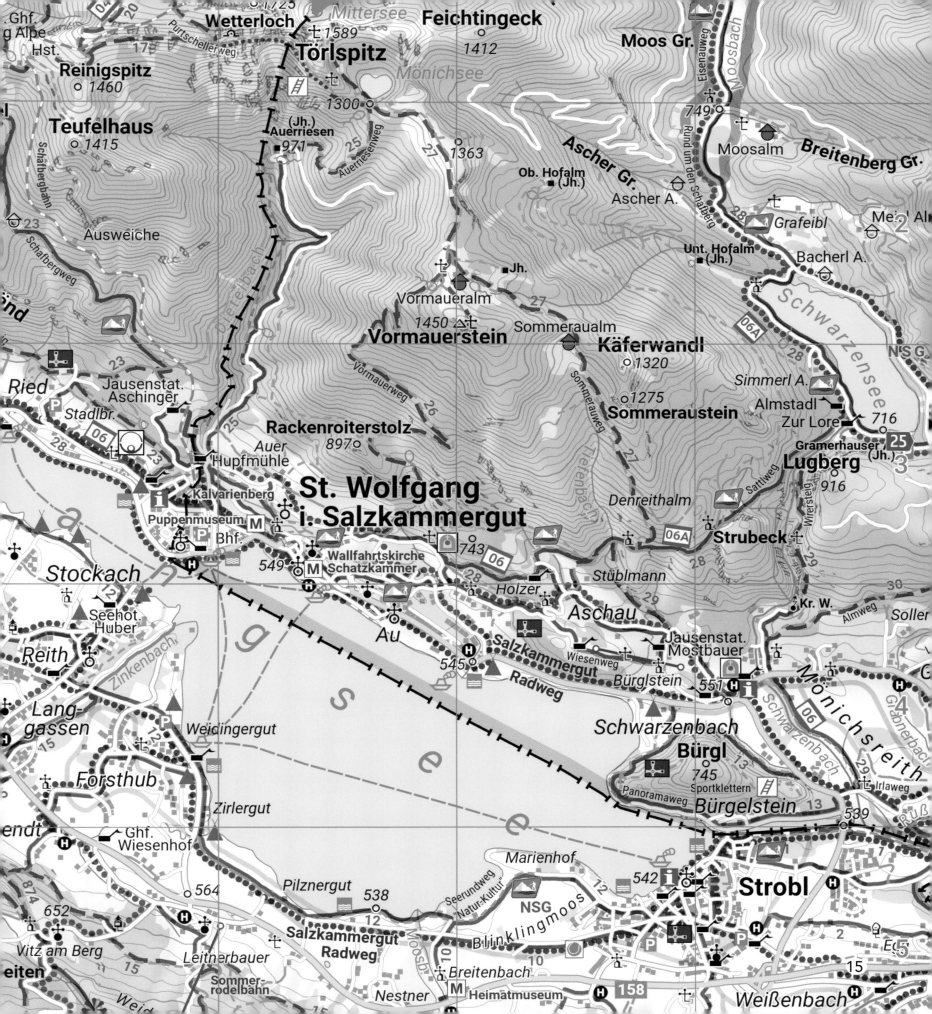

THE SALISH SEA BIOREGION

Western Washington University
Bellingham, Washington, USA
By Aquila Flower

The Salish Sea is an international estuarine ecosystem encompassing an intricate network of inland marine waterways in Washington and British Columbia. This bioregion includes both the marine waters and their upland watersheds and is home to more than 9 million people.

The concept of a cohesive Salish Sea region has become a critical focal point for local bioregional education, research, restoration, conservation, and policy development. This reference map shows topography, bathymetry, major urban areas, and water bodies and was created by harmonizing publicly available Canadian and US datasets as part of the Salish Sea Atlas.

The Salish Sea is considered an estuarine system because of the large inputs of freshwater it receives throughout the year. Streams deliver freshwater along with sediments, nutrients, organic matter, and pollutants to the sea. The watersheds of the Salish Sea bioregion can be seen in the context of the other major North American watersheds.

CONTACT
Aquila Flower
aquila.flower@wwu.edu

SOFTWARE
ArcGIS Pro, ArcGIS Spatial Analyst

DATA SOURCES
Salish Sea Atlas, NOAA, USGS, NASA, Natural Resources Canada, Natural Earth, Commission for Environmental Cooperation; Salish Sea Atlas, Natural Earth, Commission for Environmental Cooperation

Courtesy of Western Washington University.

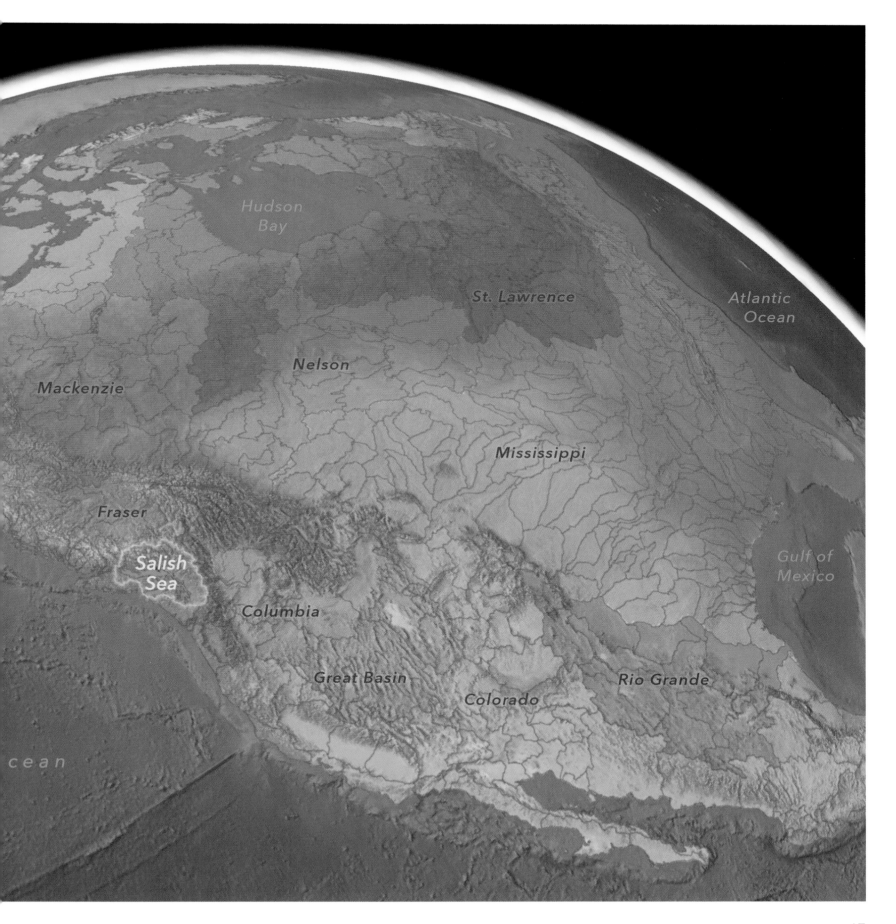

Hudson
Bay

St. Lawrence

Atlantic
Ocean

Nelson

Mackenzie

Mississippi

Fraser

Salish
Sea

Gulf of
Mexico

Columbia

cean

Great Basin

Rio Grande

Colorado

JORDANIAN AERONAUTICAL CHARTS

AED Aero Inc.
Lexington Park, Maryland, USA
By Ashley Luton

The Jordan VFR Sectional is an aeronautical chart produced by AED Aero used for air navigation under visual flight rules. It is designed to increase safety and mission effectiveness in flight operations. This advanced aviation chart is produced at the sectional scale (1:500K) and covers the Kingdom of Jordan and surrounding areas. Aeronautical, human, and basemap features are among the elements included on the chart. Some of the features include customized annotation and masking techniques, which improve the chart's overall effectiveness and aesthetic quality.

CONTACT
Ashley Luton
Ashley.Luton@aed-llc.com

SOFTWARE
ArcGIS Desktop, ArcGIS Aviation Charting

DATA SOURCES
Internally developed proprietary data

Courtesy of AED Aero.

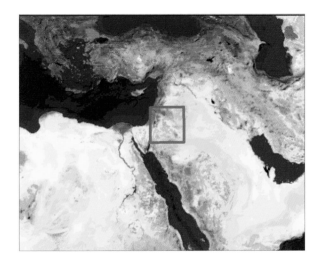

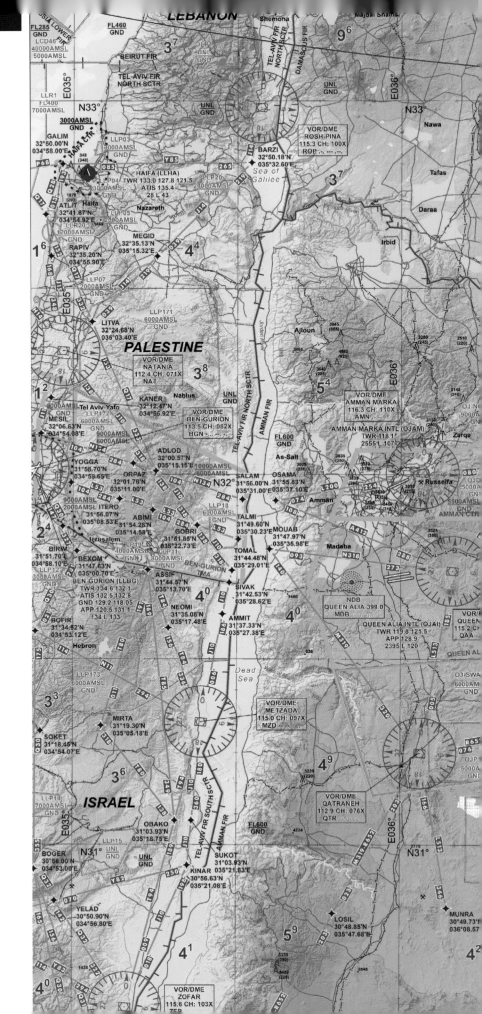

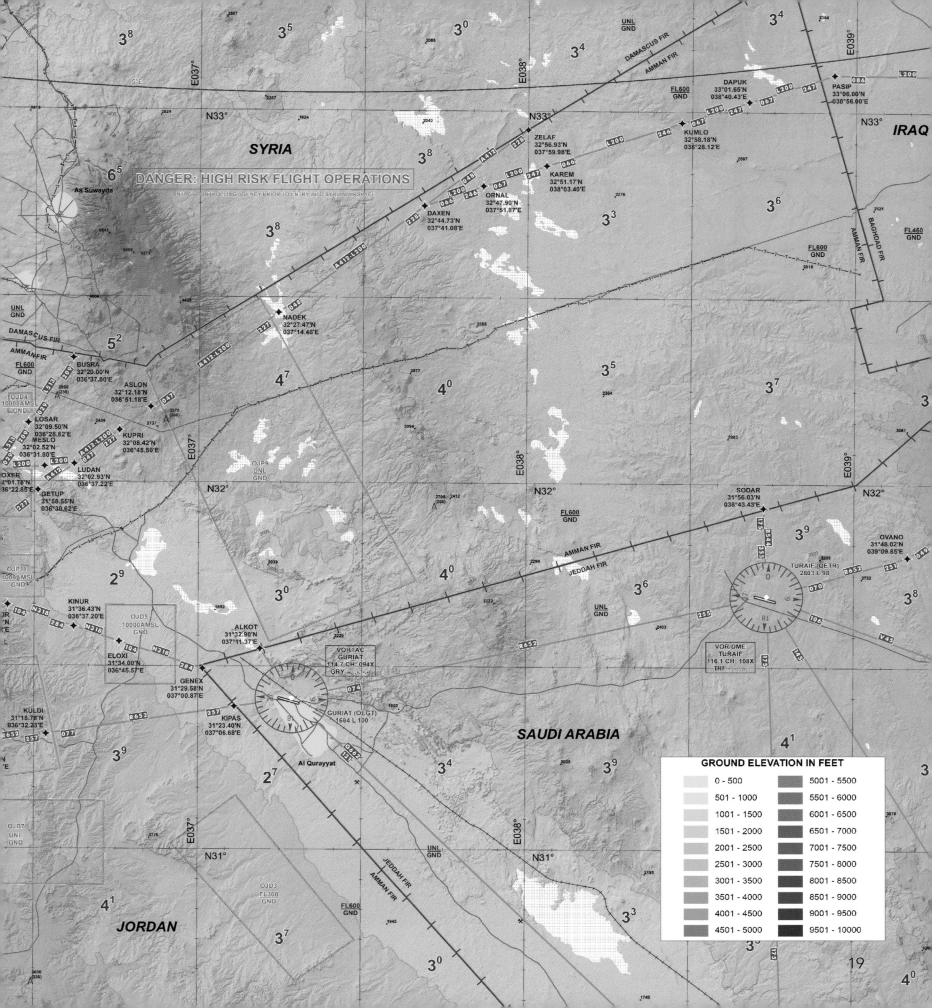

THE HALF-EARTH PROJECT

E. O. Wilson Biodiversity Foundation
Durham, North Carolina, USA
By Joel R. Johnson (E. O. Wilson Biodiversity Foundation),
D. Scott Rinnan (Yale University), Greta C. Vega,
and Estefanía Casal (Vizzuality)

The Half-Earth Project is on a mission to comprehensively map the geospatial location and distribution of the planet's species at a high enough resolution to drive decision-making. The Half-Earth Project's comprehensive map both tracks and informs conservation efforts to ensure that no species is driven to extinction from a lack of knowledge.

Data collected by scientists, satellites, and sensors paints an increasingly detailed picture of our planet. The huge amount of data on the distribution of species, human encroachment, and existing conservation areas is combined using ArcGIS API for JavaScript™. The map shows species' rarity, richness, and degree of protection. Vertebrate species are currently mapped, with maps of invertebrates, insects, and fish forthcoming. Recently released National Report Cards summarize various aspects of conservation efforts at the national level. These can be used to explore national indicators measuring conservation needs and progress and to understand the different challenges faced by each country.

CONTACT
Joel Johnson
jjohnson@eowilsonfoundation.org

SOFTWARE
ArcGIS Earth, ArcGIS Pro,
ArcGIS API for JavaScript

DATA SOURCES
Yale University, World Database on Protected Areas, World Database on Other Effective Area-Based Conservation Measures.

Courtesy of E. O. Wilson Biodiversity Foundation.

The Half-Earth Project Map
Explore biodiversity protection:

GLOBAL VERTEBRATE SPECIES

THE GLOBAL CURRENT PROTECTION

THE GLOBAL PROTECTION NEEDED

half-earth project

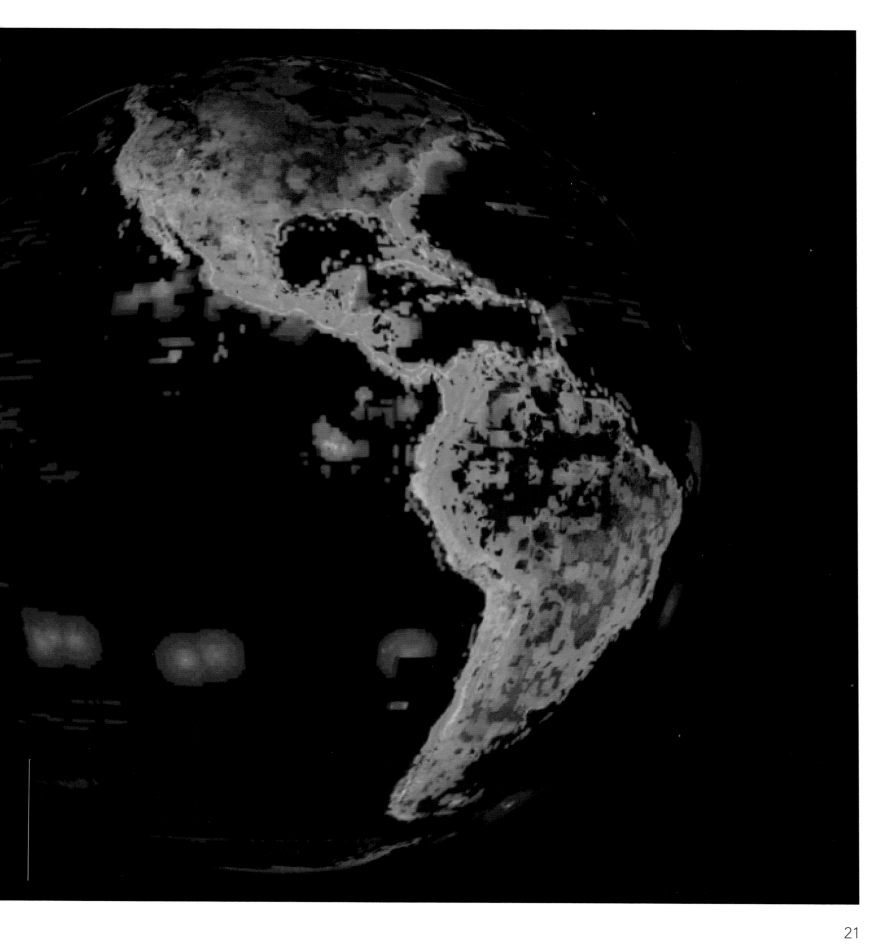

TREES AND PRIVILEGE

GreenInfo Network
Oakland, California, USA
By Kimberly Becerril

The inequality between neighborhoods in many US cities transcends income. Low-income neighborhoods disproportionately have fewer trees and hotter streets than their higher income neighbors. Trees are critical to cooling urban areas and providing shady relief from the sun. As climate change continues to cause record-breaking heat waves, building cities with an equitable tree infrastructure is essential for the well-being of all communities.

CONTACT
Kimberly Becerril
becerrilkim@gmail.com

SOFTWARE
ArcGIS Pro, Adobe Illustrator

DATA SOURCES
NASA and USGS thermal satellite imagery analyzed by NPR, USCB income, NLCD tree cover

Courtesy of Kimberly Becerril.

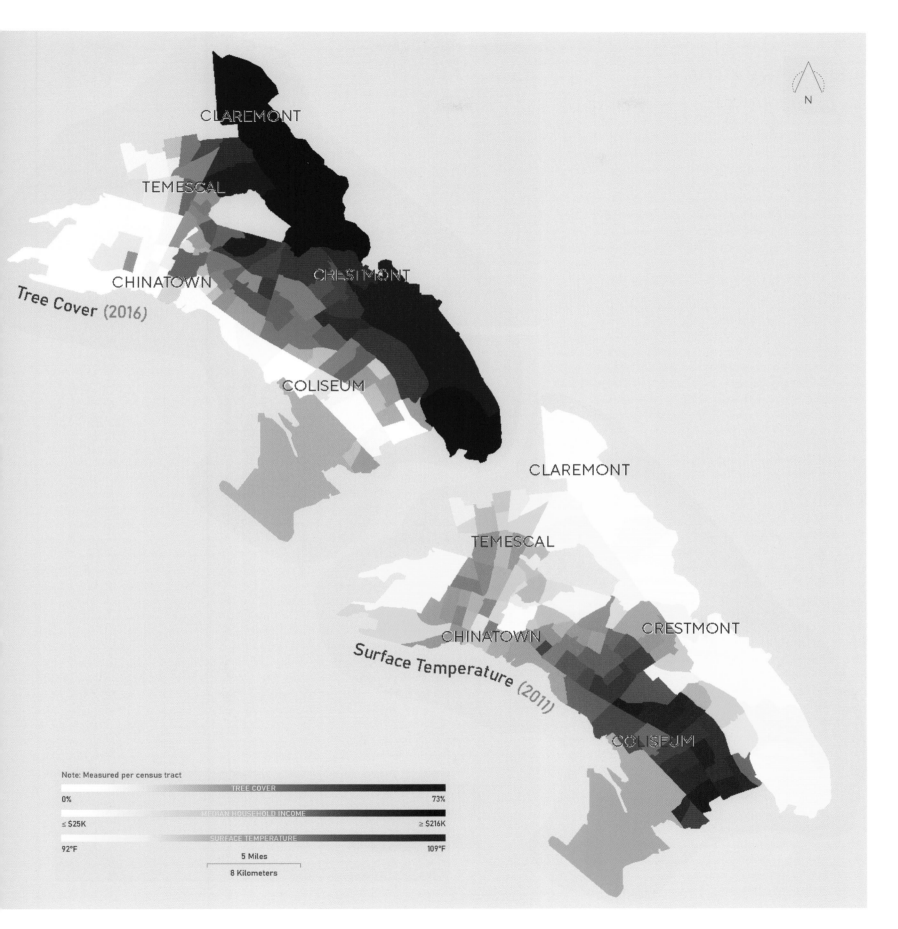

CLAREMONT

TEMESCAL

CHINATOWN

CRESTMONT

Tree Cover (2016)

COLISEUM

CLAREMONT

TEMESCAL

CHINATOWN

CRESTMONT

Surface Temperature (2011)

COLISEUM

Note: Measured per census tract

TREE COVER
0%   73%

MEDIAN HOUSEHOLD INCOME
≤ $25K   ≥ $216K

SURFACE TEMPERATURE
92°F   109°F

5 Miles
8 Kilometers

N

NEW YORK CITY: AMERICA'S MELTING POT

Tyler Morton
Broomfield, Colorado, USA
By Tyler Morton

Popularized in 1908 by Israel Zangwill, the term *melting pot* became a metaphor for a society in which nationalities, cultures, and ethnicities blended as one. Nowhere in America is this better represented than New York City. This map displays the resulting racial diversity as collected by 2015 US Census Bureau data. Symbolized using multivariate dot density, each dot represents 100 persons of a given background. The primary map displays the five main racial categories (White, Hispanic, Black, Asian, and Other) together, while the inset maps allow each category to be viewed individually. The dots offer a way to visualize the intensity and dispersion of people's self-identified racial origins or sociocultural groups.

CONTACT
Tyler Morton
tyler@inkedlandco.com

SOFTWARE
ArcGIS Pro

DATA SOURCES
Esri, US Census Bureau (2015)

Courtesy of Tyler Morton.

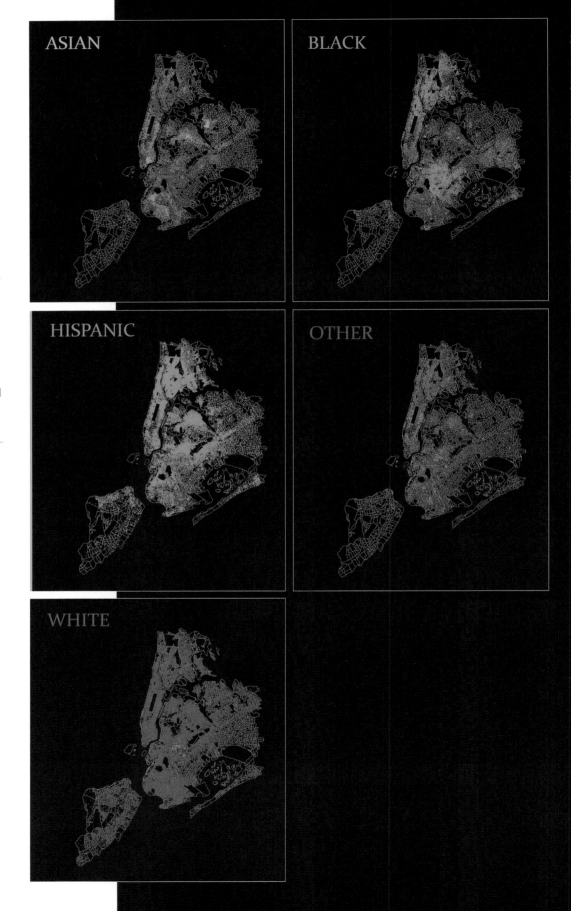

1 Dot = 100 Persons
- WHITE
- HISPANIC
- BLACK
- ASIAN
- OTHER

0   2.5   5   10 Miles

(Inset maps not to scale)

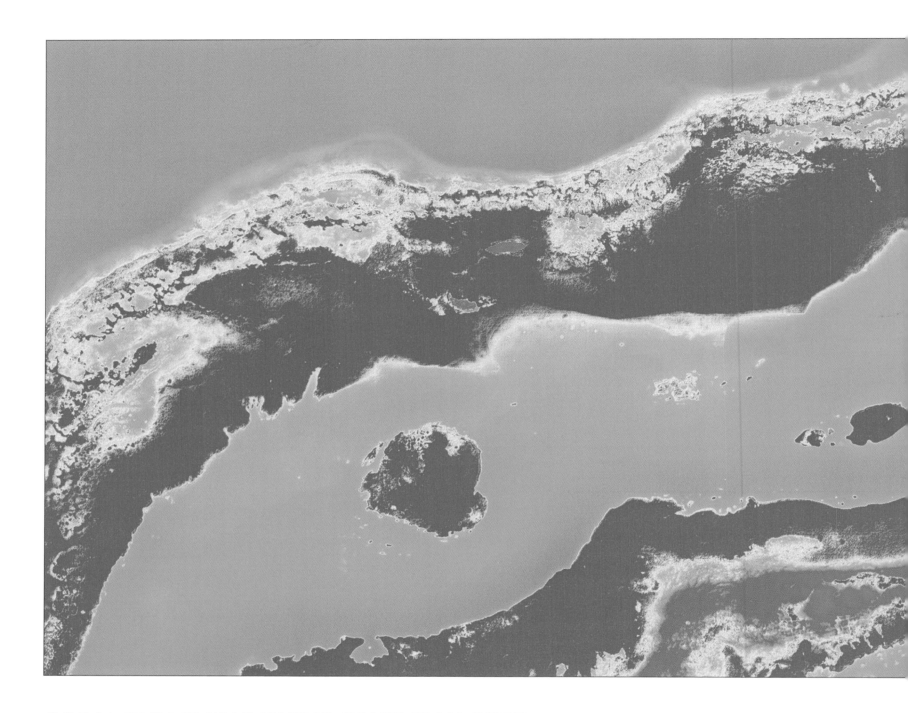

CORAL BLEACHING IN THE CALITUBAN REEF

Eötvös Loránd University
Budapest, Hungary
By Ghada Sahbeni

This image illustrates potential bleaching events in the Calituban Reef in the Philippines. The result was based on Sentinel 2-MSI data transformed using principal component analysis. This work was done as a voluntary project conducted by the United Nations Development Programme in Turkey through a Sustainable Development Goals–Artificial Intelligence (SDG-AI) Lab initiative to map coral ecosystems and assess the effects of climate change on life underwater.

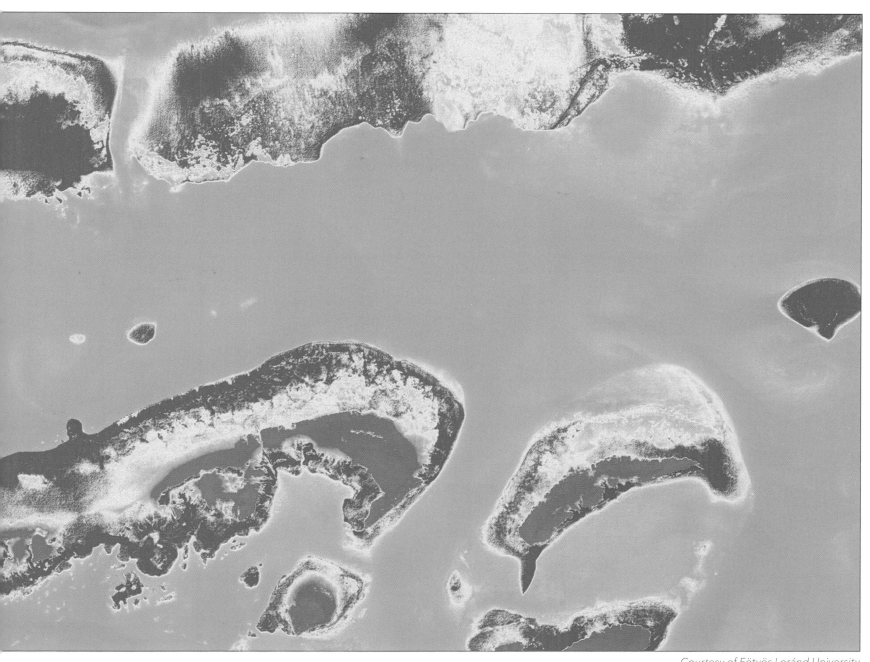

Courtesy of Eötvös Loránd University.

CONTACT
Ghada Sahbeni
gsahbeni@caesar.elte.hu

SOFTWARE
ArcGIS Desktop, ArcGIS
Spatial Analyst

DATA SOURCES
Sentinel 2

NATURAL INFRASTRUCTURE IN THE COLOMBIAN ORINOQUIA REGION

WWF Colombia
Orinoco River Basin, Colombia
By Johanna Prüssmann, Sofía Rincón, Héctor Tavera, and César Suárez

The sustainable land use (SuLu) methodology for the Colombian Orinoquia region is updated here, incorporating biodiversity, ecosystem services, legal, and regulatory considerations.

As demand for the region's main agricultural products such as palm oil, soy, rice, meat, and milk increases, pressure on the region's natural areas is growing. Additionally, climate change will most likely intensify this pressure.

As part of the International Climate Initiative's project, "Climate-smart planning in savannas, through political advocacy, management, and good practices—SuLu 2," this updated map can better guide planning processes in the region and serves as an example of an updated regional planning framework.

CONTACT
Johanna Prüssmann
jprussmann@wwf.org.co

SOFTWARE
ArcGIS Pro, ArcGIS Spatial Analyst, Adobe Illustrator

DATA SOURCES
"Estructura ecológica principal de la Orinoquia colombiana-Actualización metodológica mapa Sulu" (2020)

Courtesy of WWF Colombia.

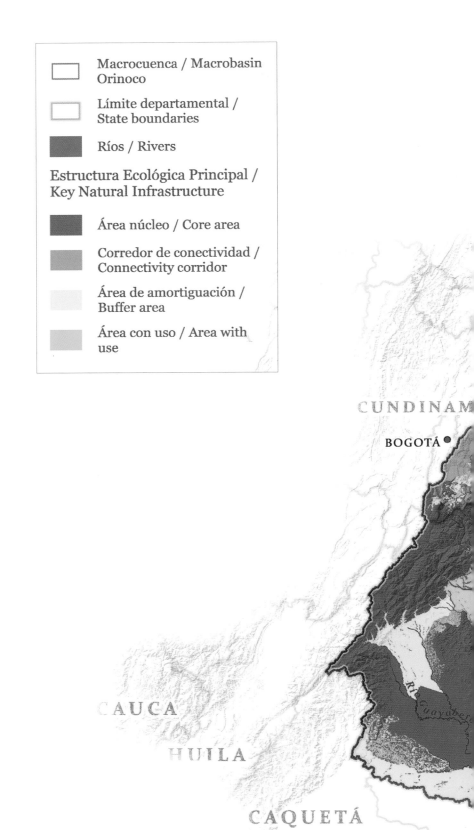

Macrocuenca / Macrobasin Orinoco

Límite departamental / State boundaries

Ríos / Rivers

Estructura Ecológica Principal / Key Natural Infrastructure

Área núcleo / Core area

Corredor de conectividad / Connectivity corridor

Área de amortiguación / Buffer area

Área con uso / Area with use

CUNDINAM

BOGOTÁ

CAUCA

HUILA

CAQUETÁ

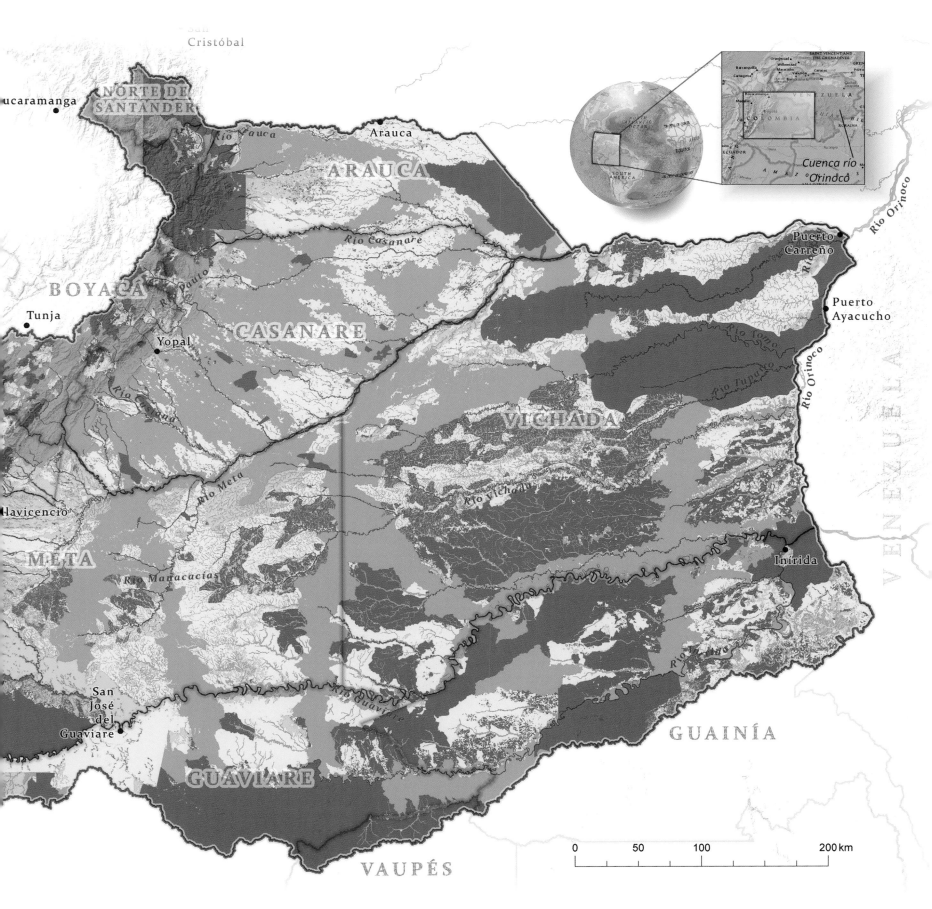

COMBATING HUMAN–ELEPHANT CONFLICT IN KAFUE NATIONAL PARK

Game Rangers International
Kafue National Park, Zambia
By Vincent Abere

Because of their proximity to Kafue National Park and known wildlife habitats and corridors, the communities bordering the Nkala Game Management Area have continued to experience high levels of conflict with wildlife and, most devastatingly, with elephants. Game Rangers International (GRI) has partnered with the International Elephant Foundation (IEF) and the Department of National Parks and Wildlife (DNPW) to equip a Human Elephant Conflict (HEC) Ranger Unit with chili guns as a tactical way to implement low-cost, homemade methods of mitigation while improving effectiveness and reducing continued conflict.

The map shows conflict data captured by ranger teams using ArcGIS Survey123 and monitored through ArcGIS Dashboards. Park managers can quickly use the map to visualize conflict hot spots and plan for ranger deployment during conflict seasons.

CONTACT
Vincent Abere
vincent@gamerangersinternational.org

SOFTWARE
ArcGIS Pro, ArcGIS Dashboards,
ArcGIS Survey123

DATA SOURCES
Game Rangers International,
Department of National Parks and Wildlife

Courtesy of Game Rangers International.

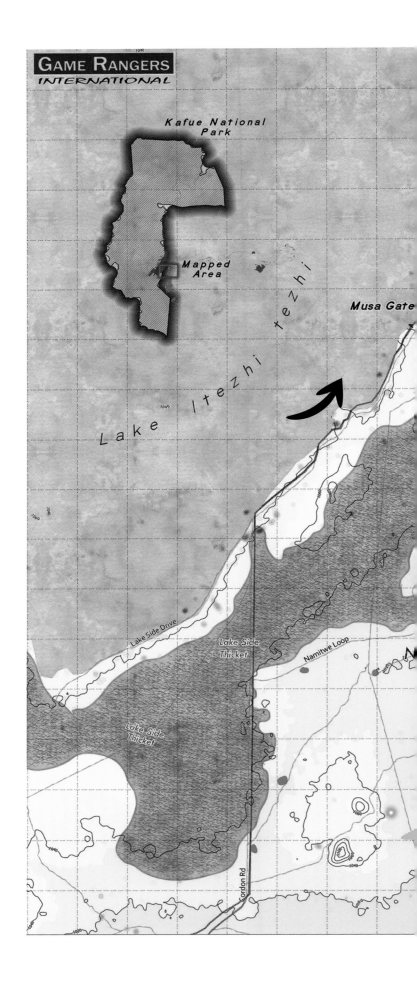

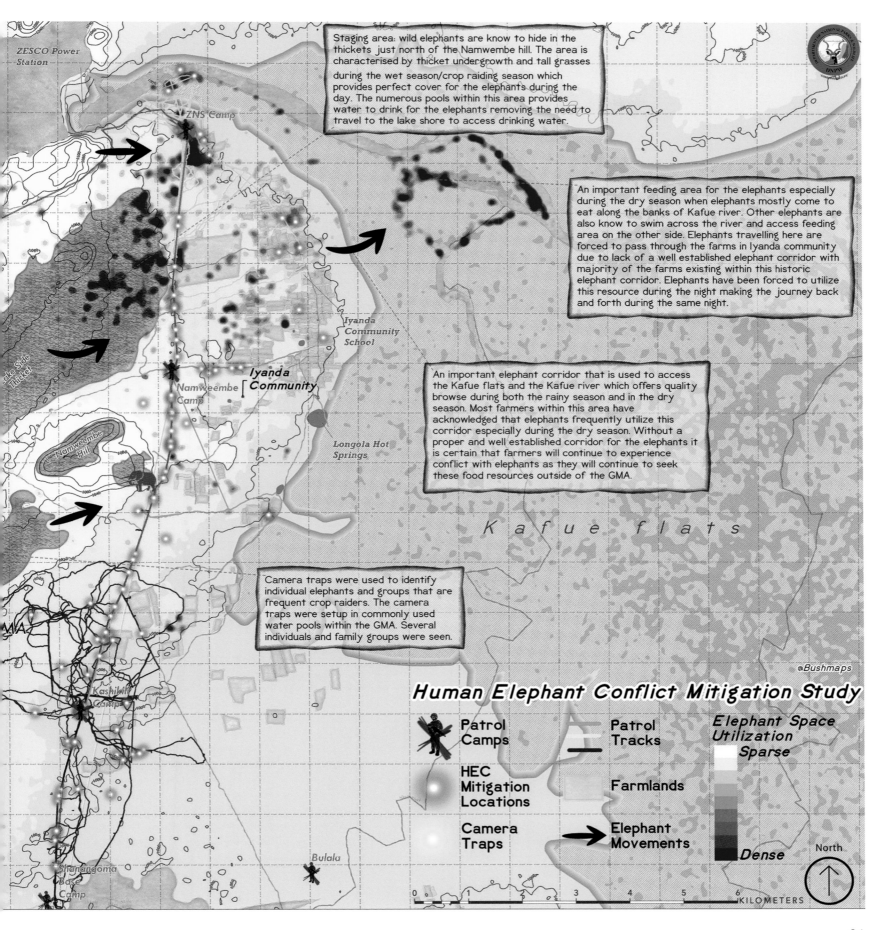

Staging area: wild elephants are know to hide in the thickets just north of the Namwembe hill. The area is characterised by thicket undergrowth and tall grasses during the wet season/crop raiding season which provides perfect cover for the elephants during the day. The numerous pools within this area provides water to drink for the elephants removing the need to travel to the lake shore to access drinking water.

An important feeding area for the elephants especially during the dry season when elephants mostly come to eat along the banks of Kafue river. Other elephants are also know to swim across the river and access feeding area on the other side. Elephants travelling here are forced to pass through the farms in Iyanda community due to lack of a well established elephant corridor with majority of the farms existing within this historic elephant corridor. Elephants have been forced to utilize this resource during the night making the journey back and forth during the same night.

An important elephant corridor that is used to access the Kafue flats and the Kafue river which offers quality browse during both the rainy season and in the dry season. Most farmers within this area have acknowledged that elephants frequently utilize this corridor especially during the dry season. Without a proper and well established corridor for the elephants it is certain that farmers will continue to experience conflict with elephants as they will continue to seek these food resources outside of the GMA.

Camera traps were used to identify individual elephants and groups that are frequent crop raiders. The camera traps were setup in commonly used water pools within the GMA. Several individuals and family groups were seen.

ZESCO Power Station

ZNS Camp

Lake Side Thicket

Iyanda Community School

Iyanda Community

Namweembe Camp

Namweembe Hill

Longola Hot Springs

Kafue flats

MA

Kashikili Camp

Bushmaps

Bulala

Shanangoma Base Camp

Human Elephant Conflict Mitigation Study

Patrol Camps

Patrol Tracks

Elephant Space Utilization

Sparse

HEC Mitigation Locations

Farmlands

Camera Traps

Elephant Movements

Dense

North

0   1   2   3   4   5   6
KILOMETERS

31

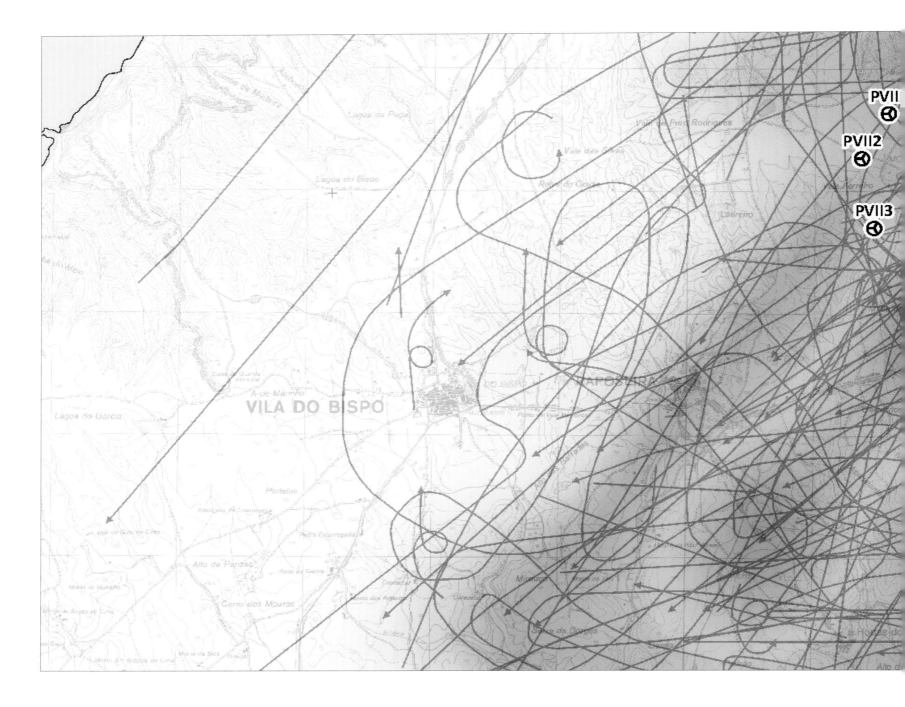

PVII

PVII2

PVII3

MIGRATION ROUTES OF THE GRIFFON VULTURE

Ecosativa
Évora, Portugal
By Joana Veríssimo

Each autumn, thousands of migratory birds fly from Portugal and other European countries to spend the winter in Africa. Griffons (*Gyps fulvus*) are one of these species. The southwest of Portugal is a common route where large flocks are visible, gliding on thermal air currents and sometimes resting before crossing the

Atlantic. Wind farms are located close to this route, so monitoring is necessary during the migration period. This requires several ornithologists, located at specific observation points, to give the alert to stop the turbines as soon as a migrating bird approaches the safety zone and thus avoid any deaths.

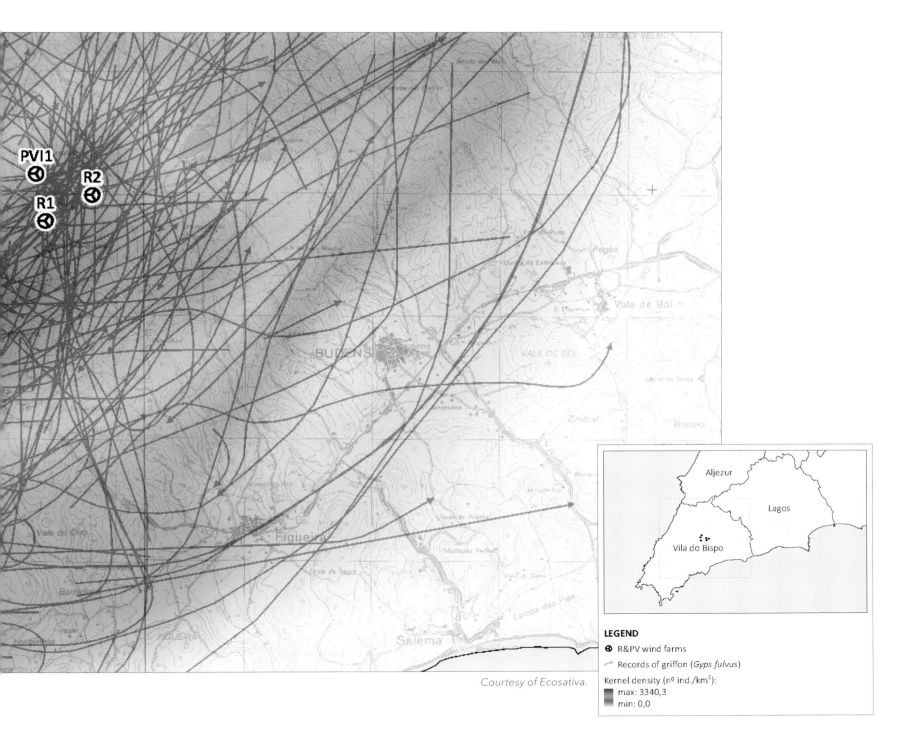

Courtesy of Ecosativa.

This map of Vila do Bispo, on the southwest coast of Portugal, is part of the annual monitoring report and illustrates the areas where the highest concentrations of griffons were recorded during the fall of 2020. The red spots indicate more than 3,300 individuals per square kilometer.

CONTACT
Joana Veríssimo
joana.c.verissimo@gmail.com

SOFTWARE
ArcGIS Desktop

DATA SOURCES
Field data from Ecosativa

ECOLOGICAL CONNECTIVITY STRATEGIES

Boffa Miskell Ltd.
Auckland, New Zealand
By Sandeep Gangar

The current extent and quality of native ecosystems in Auckland continues to decline at an alarming rate due to substantial fragmentation of important habitats. Improving ecological connectivity is a key method used to help protect biodiversity and mitigate impacts from climate change and other pressures on ecosystems. Boffa Miskell, in conjunction with Auckland Council, developed ecological connectivity strategies (ECS) for two local council boards in the Auckland region, Upper Harbour and Rodney East. Umbrella species were selected to represent a range of terrestrial ecosystem types present in each area. Modeling the connectivity of the selected species between the core habitats provided a novel insight into the dispersal patterns of the species. These maps show the connectivity patterns of the selected umbrella species of Pīwakawaka (New Zealand fantail) and Kererū (New Zealand wood pigeon).

Presenting the ECS online using the ArcGIS® StoryMaps℠ app facilitates widespread engagement in conservation activities and allows users to explore connectivity and potential conservation actions in their own area of interest.

CONTACT
Sandeep Gangar
sandeep.gangar@boffamiskell.co.nz

SOFTWARE
ArcGIS Pro, ArcGIS StoryMaps, Linkage Mapper

DATA SOURCES
Auckland Council, LINZ, Landcare Research

Courtesy of Boffa Miskell Ltd.

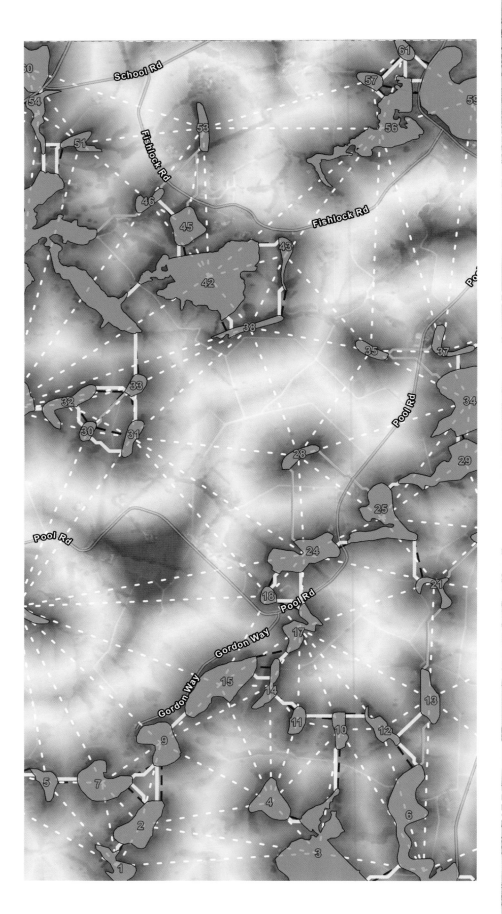

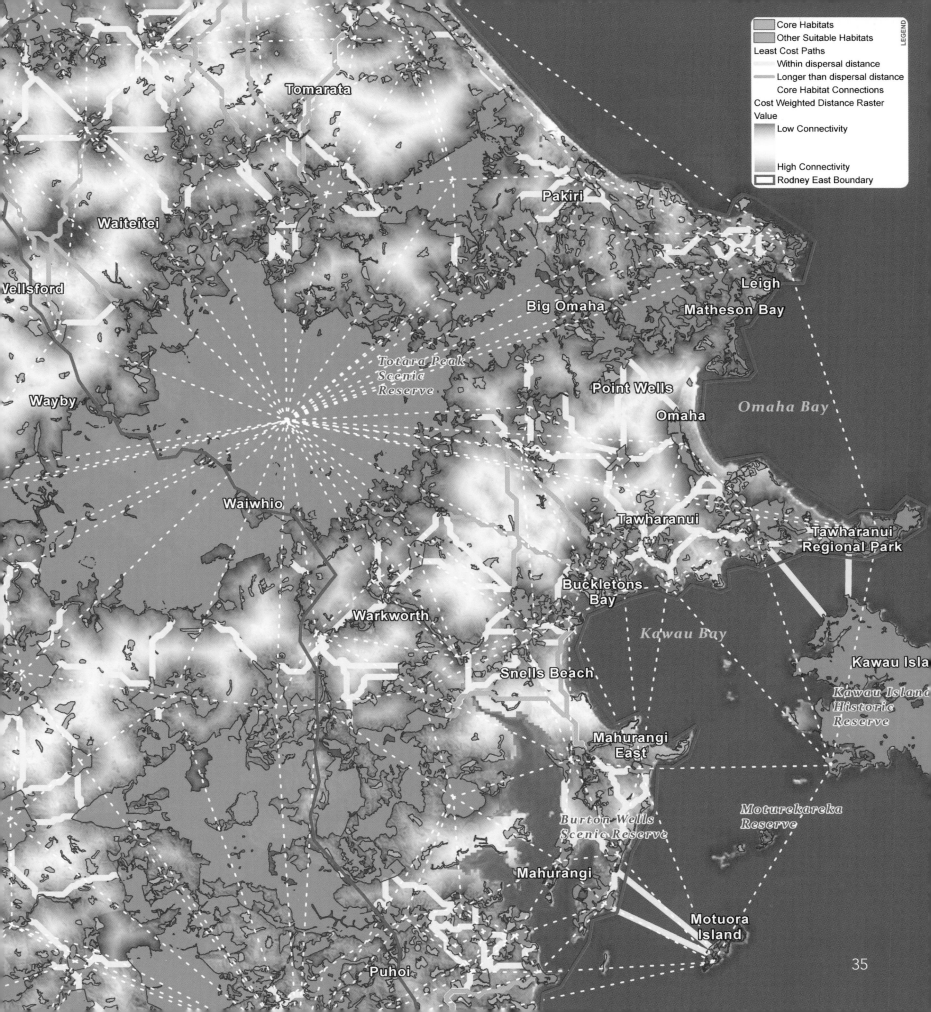

LARGE-SCALE RESTORATION OF PUBLIC LANDS ON THE COLORADO PLATEAU

Grand Canyon Trust
Utah, USA
By Stephanie Smith

From stunning petroglyphs and cliff dwellings in Bears Ears to desert vistas and dinosaur fossils in Grand Staircase-Escalante, national monuments on the Colorado Plateau protect heritage, artifacts, history, and resources.

In 2017, the Trump administration slashed Bears Ears and Grand Staircase-Escalante National Monuments by 85 percent and 47 percent respectively. These cuts opened about 2 million acres of fragile monument lands to increased resource extraction, development, and unchecked recreation. For Bears Ears it also meant robbing tribes of the opportunity to be at the forefront of stewardship of their ancestral lands in a monument they had helped create. For Grand Staircase-Escalante it meant ignoring decades of science and opening desert lands to new mining endeavors, increased cattle grazing, and destructive off-road vehicles.

The Grand Canyon Trust sees these lands as vast, interconnected, living cultural landscapes that deserve the protections a national monument provides. This map offers hope that all of Bears Ears and Grand Staircase-Escalante will be healthy and safeguarded for generations to come.

CONTACT
Stephanie Smith
ssmith@grandcanyontrust.org

SOFTWARE
ArcGIS Pro, ArcGIS® Maps for Adobe® Creative Cloud®

DATA SOURCES
Grand Canyon Trust, US Census, US Forest Service, Esri, NASA, NGA, USGS, Bureau of Land Management, National Park Service

Courtesy of Grand Canyon Trust.

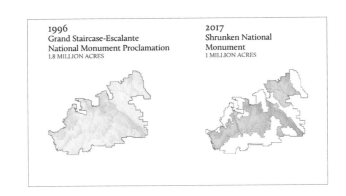

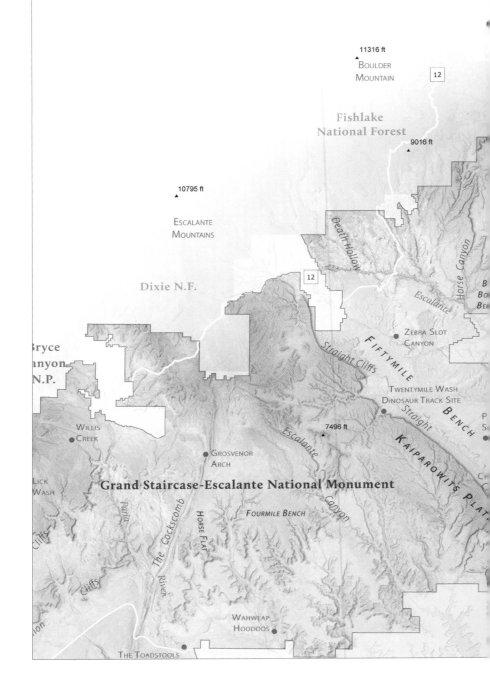

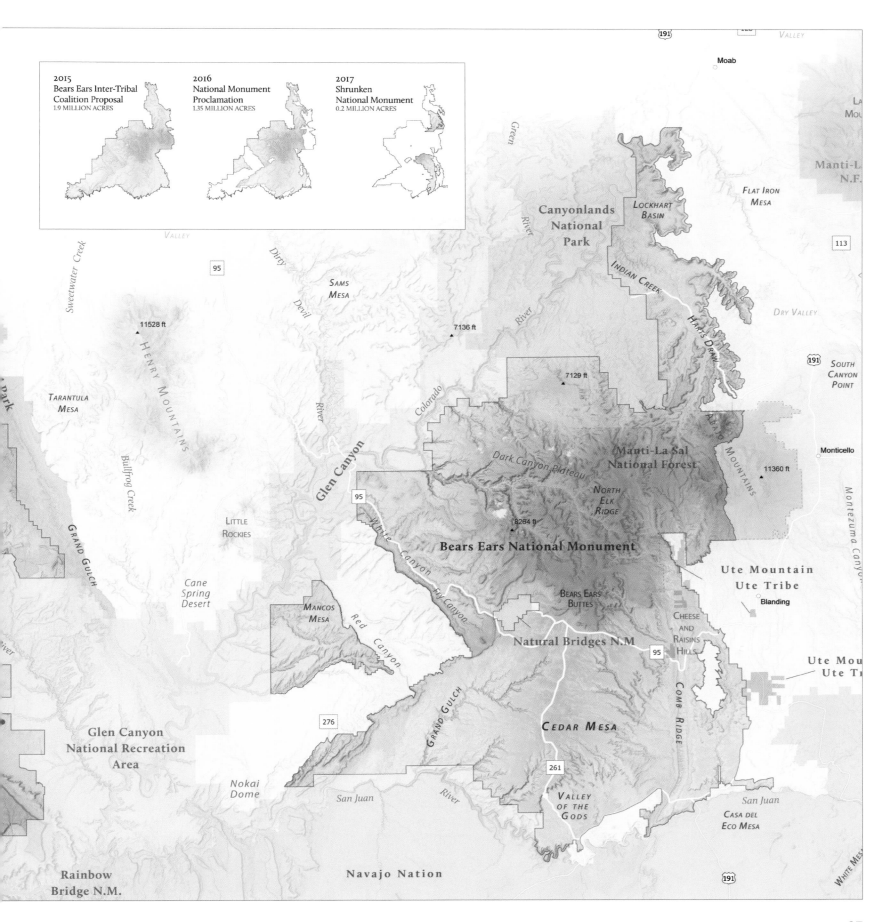

2015
Bears Ears Inter-Tribal
Coalition Proposal
1.9 MILLION ACRES

2016
National Monument
Proclamation
1.35 MILLION ACRES

2017
Shrunken
National Monument
0.2 MILLION ACRES

Moab

Canyonlands
National
Park

LOCKHART
BASIN

FLAT IRON
MESA

Manti-L
N.F.

La
Mou

113

Green

River

Devil

SAMS
MESA

11528 ft

7136 ft

7129 ft

HART'S DRAW

DRY VALLEY

191

SOUTH
CANYON
POINT

HENRY MOUNTAINS

Colorado

River

Glen Canyon

Dark Canyon Plateau

Manti-La Sal
National Forest

ABAJO MOUNTAINS

Monticello

11360 ft

Montezuma Canyon

TARANTULA
MESA

Bullfrog Creek

95

NORTH
ELK
RIDGE

LITTLE
ROCKIES

8264 ft

Bears Ears National Monument

Ute Mountain
Ute Tribe

Cane
Spring
Desert

MANCOS
MESA

Red

Canyon

BEARS EARS
BUTTES

Blanding

GRAND GULCH

White

Canyon

Fry Canyon

Natural Bridges N.M

95

CHEESE
AND
RAISINS
HILLS

Ute Mou
Ute Tr

Sweetwater Creek

VALLEY

95

Dirty

Park

GRAND GULCH

276

CEDAR MESA

COMB RIDGE

Glen Canyon
National Recreation
Area

261

Nokai
Dome

San Juan

River

VALLEY
OF THE
GODS

San Juan

CASA DEL
ECO MESA

WHITE MES

Rainbow
Bridge N.M.

Navajo Nation

191

37

MAPPING NUTRIENT AND WATER AVAILABILITY FOR DANISH FOREST PLANNING

Aarhus University, Department of Agroecology
Tjele, Denmark
By Mette Balslev Greve and Yannik Elo Roell

Denmark is a highly farmed country, but the government intends to double forest cover within 80 to 100 years from a 1989 baseline. The aim is to create productive, healthy, stable, and biodiverse forests. When deciding which tree species to plant while afforesting or reforesting a site, it is important to know the water and nutrient status of the soil, so that the tree species will thrive in a particular area. For example, tree species such as Sitka spruce will grow in very dry areas that also are low in nutrients, whereas European ash requires higher amounts of nutrients and will grow in very wet areas.

For this map of nutrient and water supply in Denmark, the landscape was classified into six nutrient classes and nine water classes based on four variables: pH at 1–2 m depth, average precipitation between April and October, groundwater depth, and plant-available water. With this map, it is now possible to identify which species will thrive in an area based on water and nutrient availability.

CONTACT
Mette Balslev Greve
metteb.greve@agro.au.dk

SOFTWARE
ArcGIS Pro

DATA SOURCES
Department of Agroecology, Aarhus University

Courtesy of Aarhus University, Department of Agroecology.

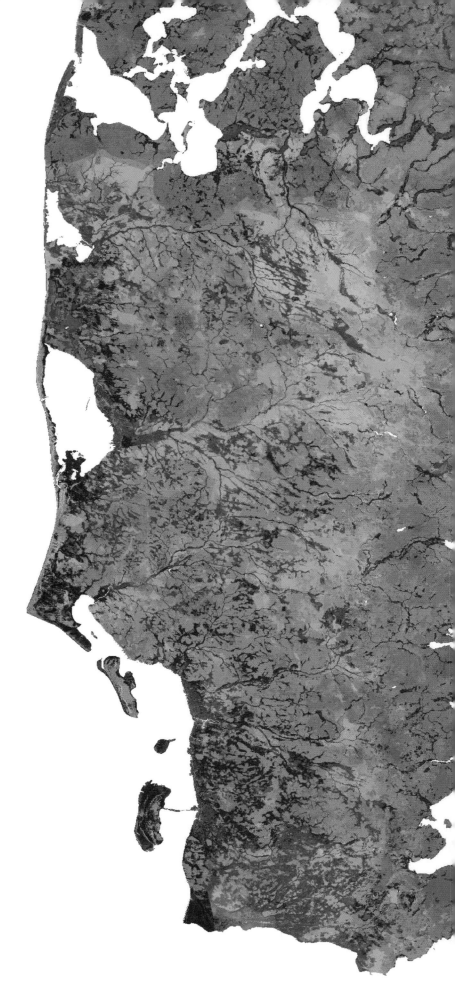

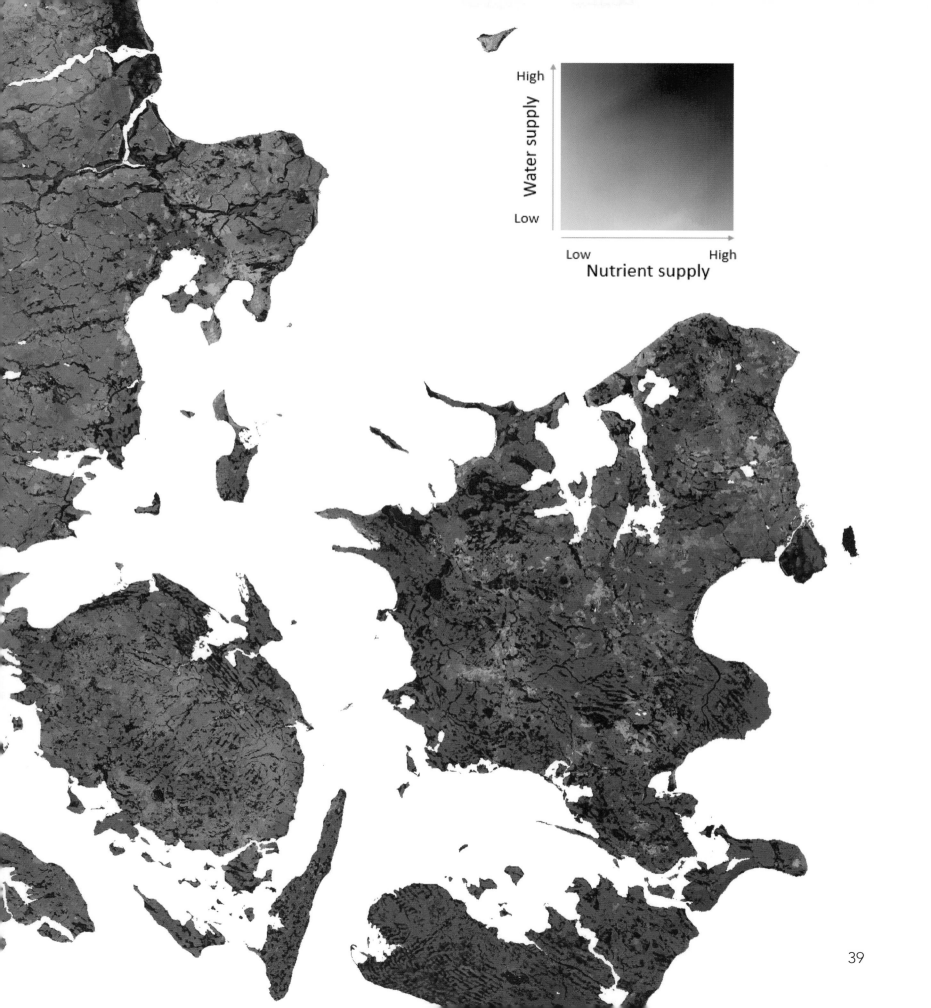

MANGROVE AND POND CHANGES IN THE MAHAKAM DELTA, INDONESIA, 1990–2019

Yayasan Konservasi Alam Nusantara
Kutai Kartanegara, East Kalimantan, Indonesia
By Dzimar Prakoso

This map shows the changes in the presence of mangroves in the Mahakam Delta, Indonesia, over time. Three snapshots of pond land cover reveal the mangrove losses from 1990 to 2019 and demonstrate a correlation with the greater region during the same period.

CONTACT
Dzimar Akbarur Rokhim Prakoso
Dzimar.prakoso@ykan.or.id

SOFTWARE
ArcGIS Pro, ArcGIS Desktop, Google Earth Engine

DATA SOURCES
Landsat 5 TM Imagery (1990), Landsat 7 +ETM (2000),
Landsat 8 OLI (2019), Fieldwork data in East Kalimantan (2019),
Coastline Data from BIG (2017),
Administrative Boundary from BIG (2017)

Courtesy of Yayasan Konservasi Alam Nusantara.

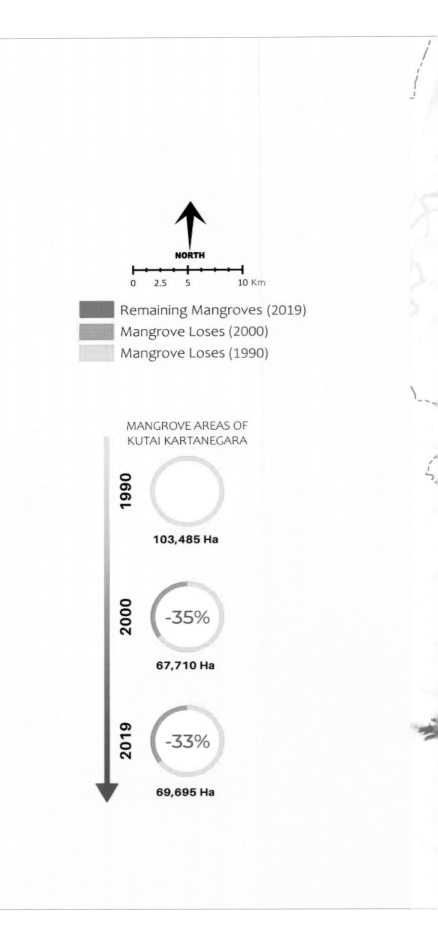

NORTH

0   2.5   5   10 Km

Remaining Mangroves (2019)
Mangrove Loses (2000)
Mangrove Loses (1990)

MANGROVE AREAS OF
KUTAI KARTANEGARA

1990

103,485 Ha

2000   -35%

67,710 Ha

2019   -33%

69,695 Ha

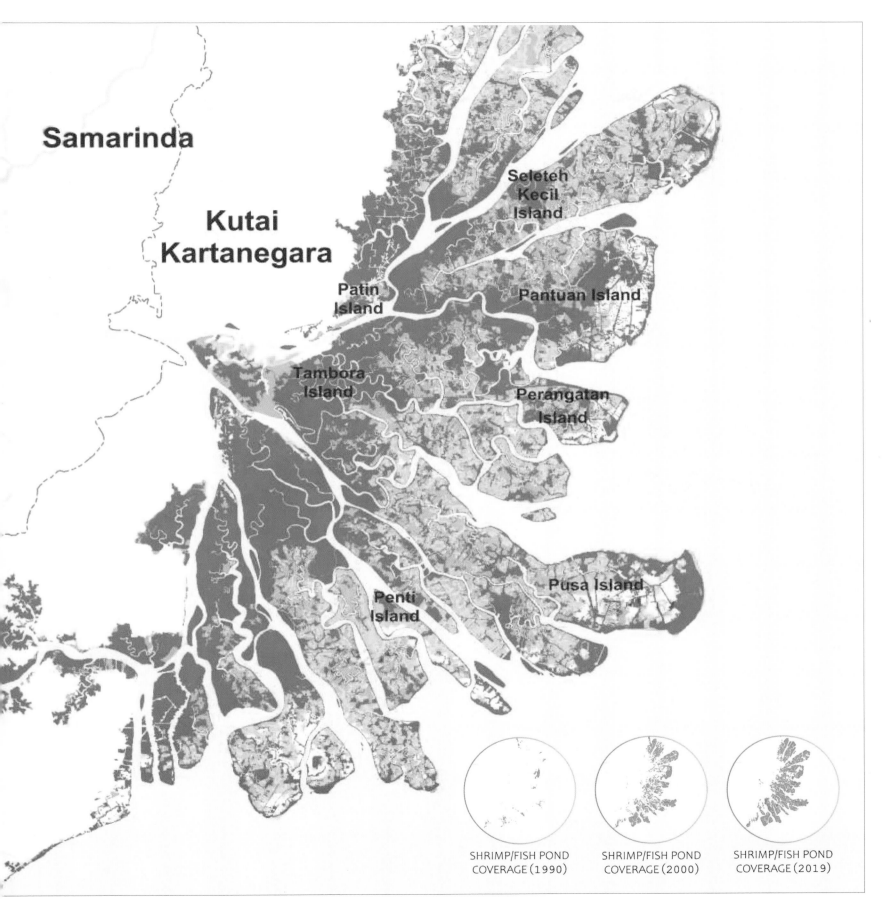

Samarinda

Kutai
Kartanegara

Seleteh
Kecil
Island

Patin
Island

Pantuan Island

Tambora
Island

Perangatan
Island

Penti
Island

Pusa Island

SHRIMP/FISH POND
COVERAGE (1990)

SHRIMP/FISH POND
COVERAGE (2000)

SHRIMP/FISH POND
COVERAGE (2019)

THE FOREST STEWARDSHIP COUNCIL GIS PORTAL

Blue Raster LLC
Arlington, Virginia, USA
By Angela Wertman

The Forest Stewardship Council (FSC) GIS Portal provides access to forest-relevant spatial information such as tree-cover loss, intact forest landscapes, protected areas, Indigenous and community lands, watercourses, up-to-date satellite imagery from various sources, and boundary data. This spatial analysis identifies, highlights, and locates the characteristics of forest cover within certified areas or areas of interest.

CONTACT
Eric Ashcroft
eashcroft@blueraster.com

SOFTWARE
ArcGIS API for JavaScript, ArcGIS Image Server

DATA SOURCES
Forest Stewardship Council, ArcGIS Living Atlas of the World, World Resources Institute, NASA, University of Maryland, WWF, numerous government and national agencies

Courtesy of Blue Raster LLC.

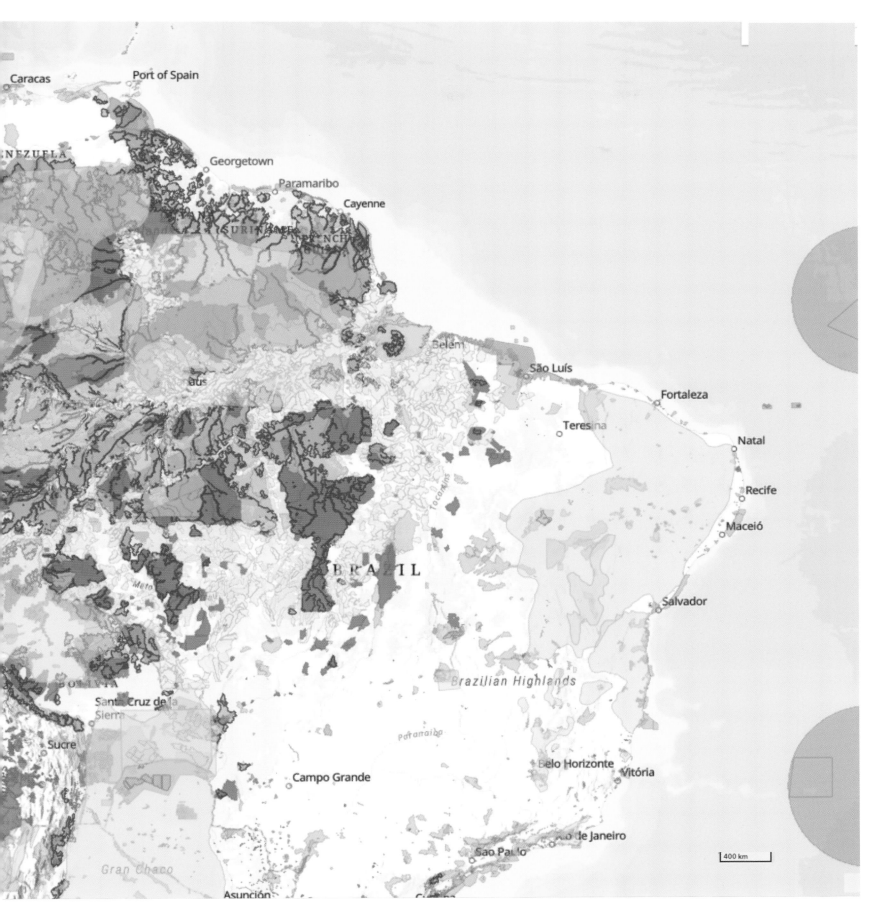

MONITORING PEATLAND RESTORATION IN INDONESIA

Greenpeace Southeast Asia,
Indonesia
By Sapta Ananda

Greenpeace Indonesia has attempted to verify the government's claims of success with peatland restoration, but it has proven difficult because of the lack of transparency and accessibility of relevant information, inconsistent figures, and the unavailability of maps of restored concession areas. In the absence of independent verification, Greenpeace decided to conduct an analysis of Critical Peat Hydrological Units in seven provinces. The goal of this analysis is to establish the locations of the degraded peatlands that have been prioritized for restoration by the government and to assess the condition of the peat landscapes to see to what extent degraded peatlands show indications of restoration through the governmental program.

Courtesy of Greenpeace Southeast Asia.

CONTACT
Sapta Ananda Proklamasi
sapta.ananda.proklamasi@greenpeace.org

SOFTWARE
ArcGIS Dashboards,
ArcGIS Pro

DATA SOURCES
KLHK, BRG, Greenpeace

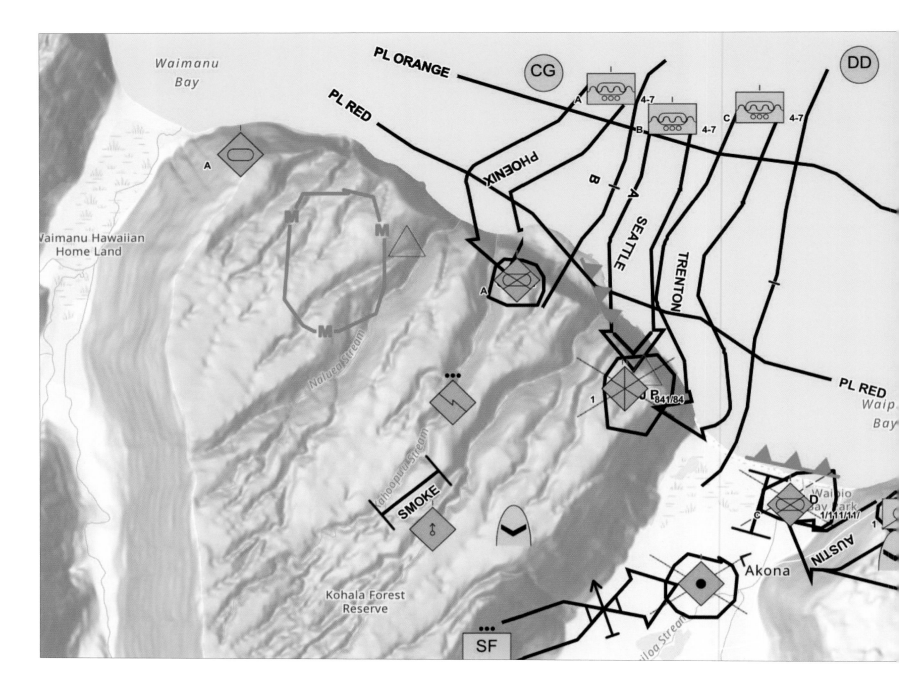

NATURAL LANGUAGE USER INTERFACES FOR MILITARY PLANNING AND AUTOMATED WAR-GAMING

Hyssos Tech LLC
Olympia, Washington, USA
By Sean Johnson

Sketch-Thru-Plan (STP) for ArcGIS Pro provides an accessible human–machine interface that avoids disruptions caused by tool complexity, providing the means to quickly generate better plans through artificial intelligence (AI) and simulation. It fuses voice and sketch input to assist military plan creation and drive simulators for automated war-gaming. The next-generation user interface enhances course of action creation and compresses the observe-orient-decide-act (OODA) loop, allowing users to plan and execute faster than their adversary.

This example of planning on a 30-foot, touch-enabled, projected wall immerses the war-gamer in the creative process of planning, focusing on the

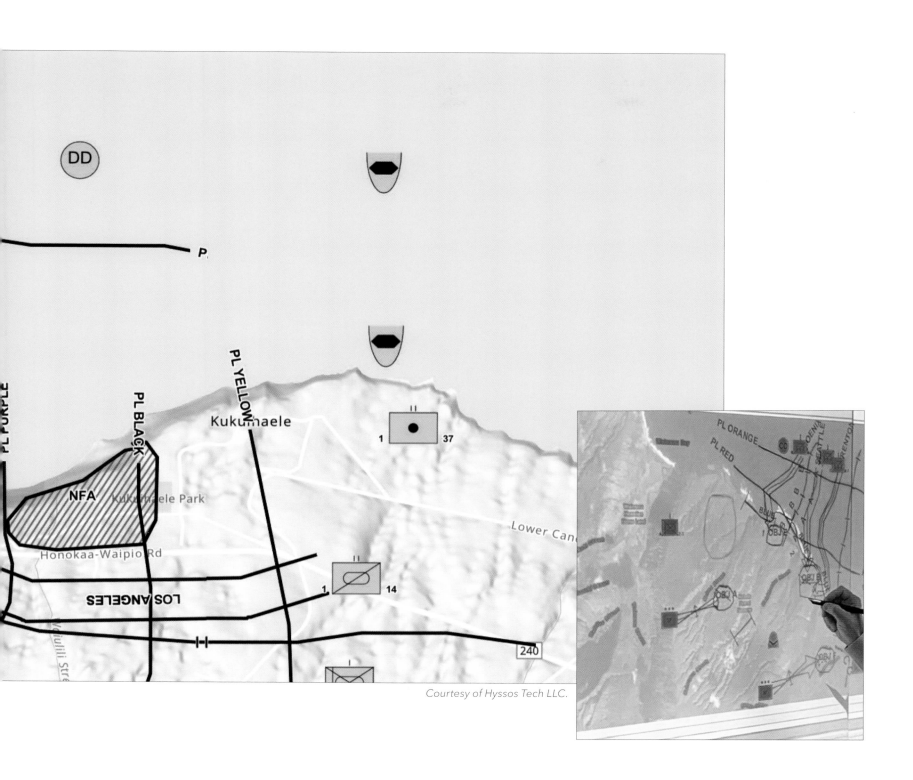

Courtesy of Hyssos Tech LLC.

task, not the tool. It applies robust AI capabilities, including natural language processing, computer vision, and computational linguistic techniques to extract meaning from combined human sketches and speech, seamlessly integrating plan outputs that drive simulators for battlefield planning.

CONTACT
Sean Johnson
Sean.Johnson@hyssos.com

SOFTWARE
ArcGIS Pro

DATA SOURCES
ArcGIS base data and on-the-fly military planning course of action

PUERTO RICO AND US VIRGIN ISLANDS TREE CANOPY

US Department of Agriculture (USDA), Puerto Rico and US Virgin Islands By Eric Rounds, Sean Patterson, Shiona Howard, Marie Schleicher, and Maya Quiñones

This map shows the results, in red, of a semiautomated process to map tree canopy cover using USGS lidar data from 2018. Canopy cover was derived using a standard height threshold of 2 meters or greater to identify a tree. Further processes included creating building footprints and removing power lines. QA/QC procedures were then used to remove non-tree data from the final canopy cover dataset.

The USDA Forest Service International Institute of Tropical Forestry and the Geospatial Technology and Applications Center have published high-resolution tree canopy cover datasets for Puerto Rico and the US Virgin Islands. These datasets will be used in urban forestry planning and green infrastructure mapping projects, including potential vegetation placement and socioeconomic impact maps.

Courtesy of US Department of Agriculture.

CONTACT
Nathan Pugh
nathan.pugh@usda.gov

SOFTWARE
ArcGIS Pro,
ArcGIS Spatial Analyst

DATA SOURCES
USGS 3DEP lidar,
USDA high-resolution orthoimagery,
Hexagon

FIRE SEVERITY MAPPING: K'GARI (FRASER ISLAND) BUSHFIRE OCTOBER-DECEMBER 2020

Department of Environment and Science
Queensland, Australia
By Marcus Toyne

K'gari (Fraser Island) is the world's largest sand island and features complex dune systems with an array of rare and unique features, including dune lakes and tall rain forests. In October 2020, an illegal campfire on the World Heritage listed site got out of control. The fire raged for more than two months and burned more than 80,000 hectares (197,000 acres)—over half the island.

This map shows the estimated severity of the fire. The map was created using a field-verified classification scheme that was applied to a delta normalized burn ratio image derived from satellite remote sensing. This analysis formed the basis for an assessment of the fire's impact on the island.

CONTACT
Marcus Toyne
marcus.toyne@daf.qld.gov.au

SOFTWARE
ArcGIS Pro

DATA SOURCES
Copernicus Sentinel data, Queensland government, ArcGIS Living Atlas of the World

Courtesy of Department of Environment and Science.

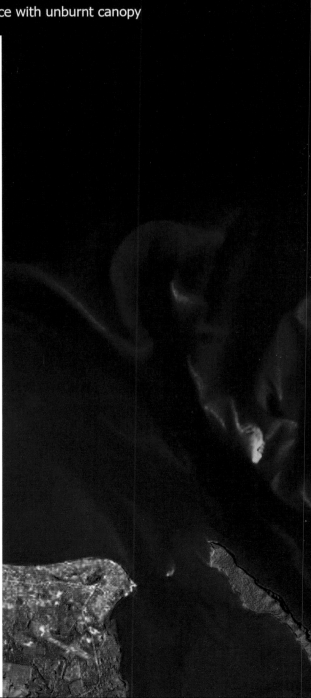

Fire Severity
- Extreme - Full canopy consumption
- High - Full canopy scorch (+/- partial canopy consumption)
- Moderate - Partial canopy scorch
- Low - Burnt surface with unburnt canopy

Platypus Bay

Orchid Beach

Happy Valley

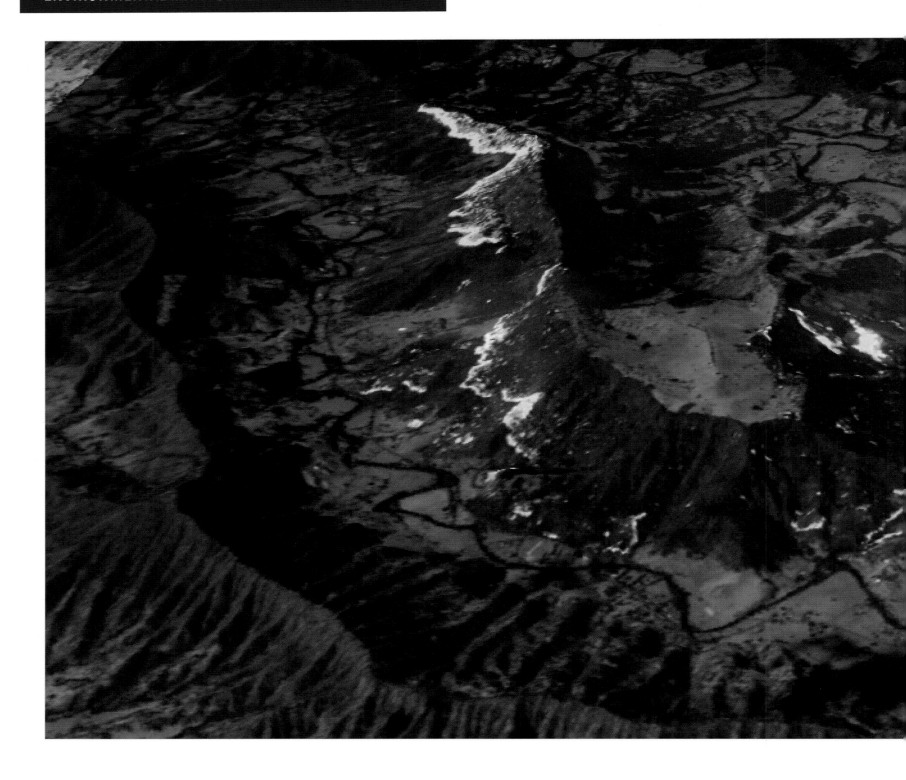

SARABAH FIRE, QUEENSLAND, 2019

Queensland Fire and Emergency
Services (QFES)
Queensland, Australia
By Lidia Dudina

This image shows the extent of the Sarabah bushfire in
3D. The fire blazed through Lamington National Park in
Queensland's Scenic Rim causing significant damage.

The image was created using two line scan images,
taken on September 4–5, blended to provide eye-
opening insight on the fire behavior. The line scan
images were taken from low-flying aircraft with a
linear array sensor.

Courtesy of QFES.

CONTACT
Lidia Dudina
Dudina.Lidia@police.qld.gov.au

SOFTWARE
ArcGIS Pro

DATA SOURCES
Esri imagery,
line scan images

STRUCTURE VULNERABILITY PREDICTIONS BEFORE A WILDFIRE EVENT

FlameMapper LLC
Agoura Hills, California,
USA
By Oliver J. Curtis, Shea
Broussard, and Anthony
Shafer

FlameMapper uses machine learning to predict the relative vulnerability of structures to a wildfire. Decades of field and modeling experience allow FlameMapper to build and evolve models that assess this risk. The model combines common topographic factors such as slope with more complex variables that evaluate structure-to-structure ignition potential.

This image is from a web app that was created for the insurance industry to assist with visualizing risk probabilities. FlameMapper chose a 3D web app for users to visually explore individual structure risk and compare it with the surrounding area. Understanding the context allows for additional exploration and discovery.

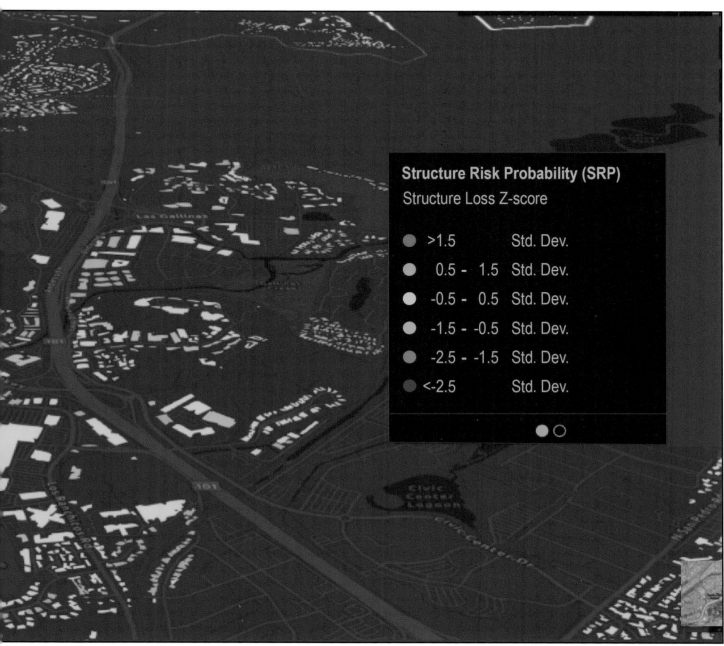

Structure Risk Probability (SRP)
Structure Loss Z-score

- >1.5   Std. Dev.
- 0.5 - 1.5   Std. Dev.
- -0.5 - 0.5   Std. Dev.
- -1.5 - -0.5   Std. Dev.
- -2.5 - -1.5   Std. Dev.
- <-2.5   Std. Dev.

Courtesy of FlameMapper LLC.

CONTACT
Oliver J. Curtis
oliver@flamemapper.com

SOFTWARE
ArcGIS Pro,
ArcGIS Experience Builder

DATA SOURCES
California Department of Forestry and
Fire Protection Damage Inspection (DINS),
FlameMapper LLC, Microsoft (ODbL),
LANDFIRE: DOI, USGS, USDA

CRACKING DOWN ON CALGARY CRIME: CRIME RATES IN 2D AND 3D

Southern Alberta Institute of Technology, Calgary, Alberta, Canada
By Dakota Tryhuba

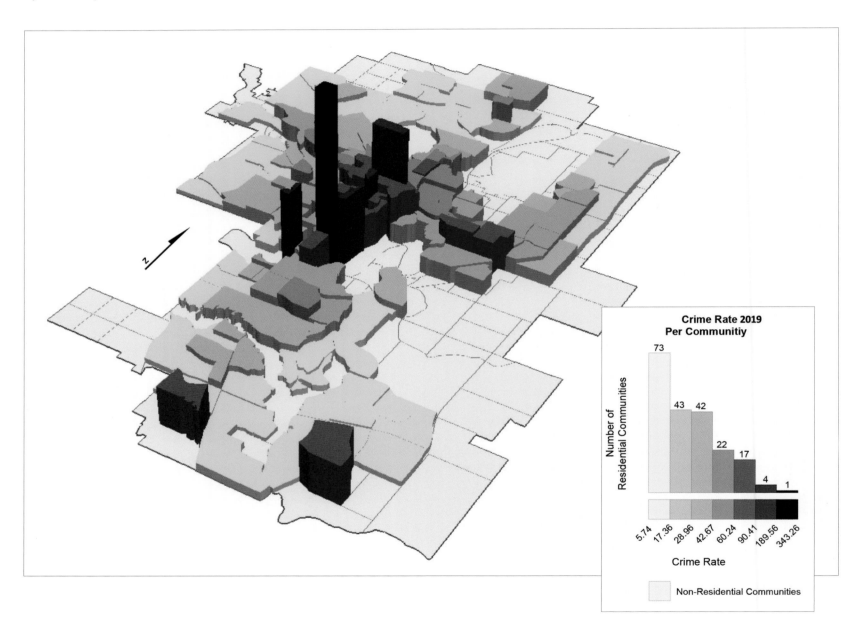

Crime mapping is used globally by analysts in law enforcement agencies to map, visualize, and analyze crime incident patterns. Using GIS, crime analysts can overlay datasets such as census demographics and reported crime incidents to help law enforcement administrators make better decisions, target resources, and formulate strategies. This data can inform tactical analysis, such as crime forecasting and geographic profiling.

The maps presented here show a 2D and 3D comparison of the residential crime rate for Calgary, Alberta, Canada. According to its Community Crime Statistics, available on the city's Open Data Portal, the crime rate in Calgary in 2019 was 24.2%, although about 59% of communities had a rate greater than the annual crime rate.

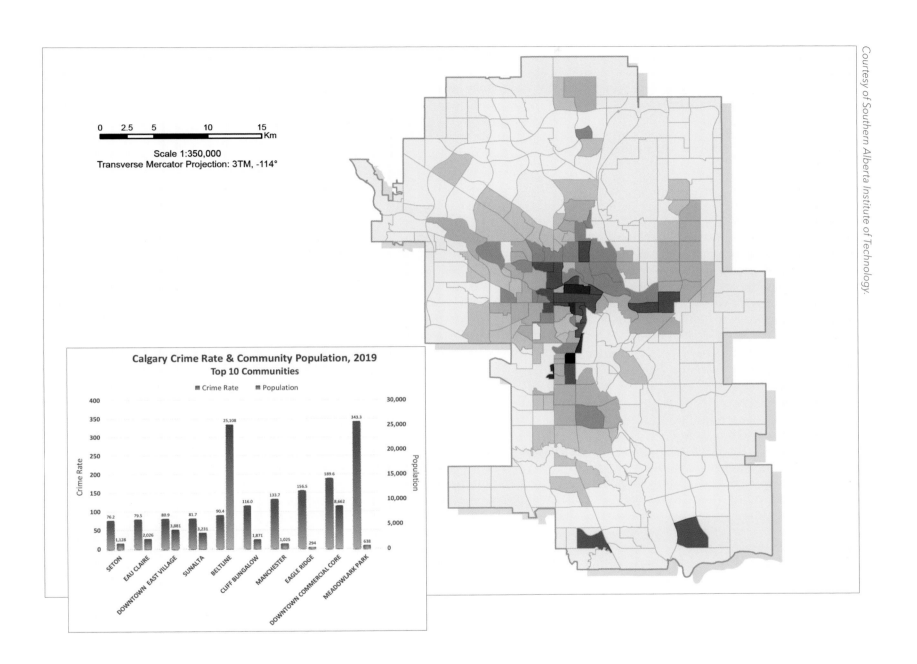

Scale 1:350,000
Transverse Mercator Projection: 3TM, -114°

Calgary Crime Rate & Community Population, 2019
Top 10 Communities

■ Crime Rate ■ Population

CONTACT
Dakota Tryhuba
dakota.tryhuba@edu.sait.ca

SOFTWARE
ArcGIS Desktop,
ArcGIS® 3D Analyst™

DATA SOURCES
Calgary Open Data,
Community Crime Statistics

THE RESILIENCE ANALYSIS AND PLANNING TOOL

Federal Emergency Management Agency (FEMA) National Integration Center
Washington, DC, USA
By Benjamin Rance

This map was created using FEMA's Resilience Analysis and Planning Tool (RAPT). RAPT is a free, no-login-required GIS tool for emergency managers to explore the intersection of population, hazards, and infrastructure. This map integrates these three aspects to improve decision-making before, during, and after disasters. Historic, real time, and some projected climate condition layers are included.

This map uses the incident analysis tool to visualize a hypothetical tornado striking downtown Houston, Texas. The orange highlighted area indicates the tornado path, the orange numbers show clusters of nursing homes near the path and the purple points show nursing homes outside the path. This tool can be used in combination with a hazard layer to highlight infrastructure at risk and provide critical details about each location. The basemap shows the percentage of the population at the census tract level who lack health insurance, and the point layer shows nursing home locations with important information for each, including the name, address, and number of beds.

CONTACT
FEMA Technical Assistance Branch
FEMA-TArequest@fema.dhs.gov

SOFTWARE
ArcGIS Online

DATA SOURCES
US Census American Community Survey,
Homeland Infrastructure Foundation-Level Data

Courtesy of FEMA National Integration Center.

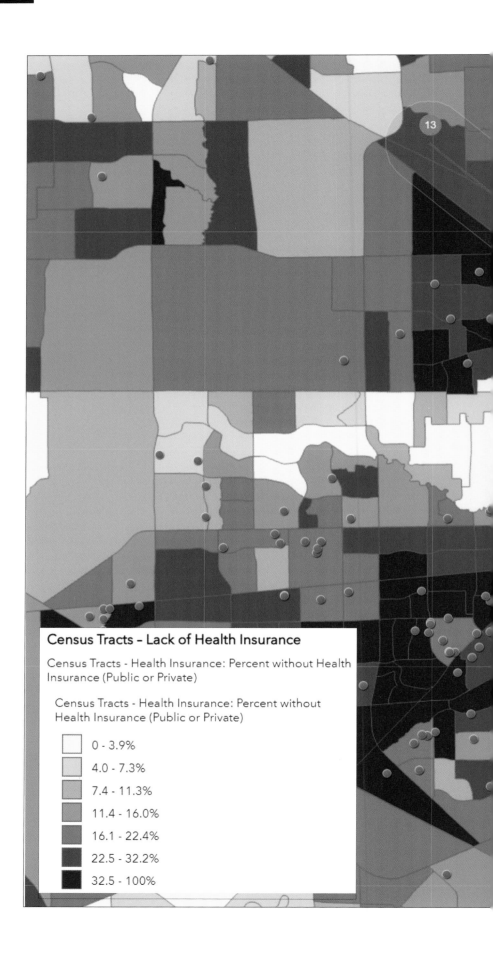

Census Tracts - Lack of Health Insurance

Census Tracts - Health Insurance: Percent without Health Insurance (Public or Private)

Census Tracts - Health Insurance: Percent without Health Insurance (Public or Private)

- 0 - 3.9%
- 4.0 - 7.3%
- 7.4 - 11.3%
- 11.4 - 16.0%
- 16.1 - 22.4%
- 22.5 - 32.2%
- 32.5 - 100%

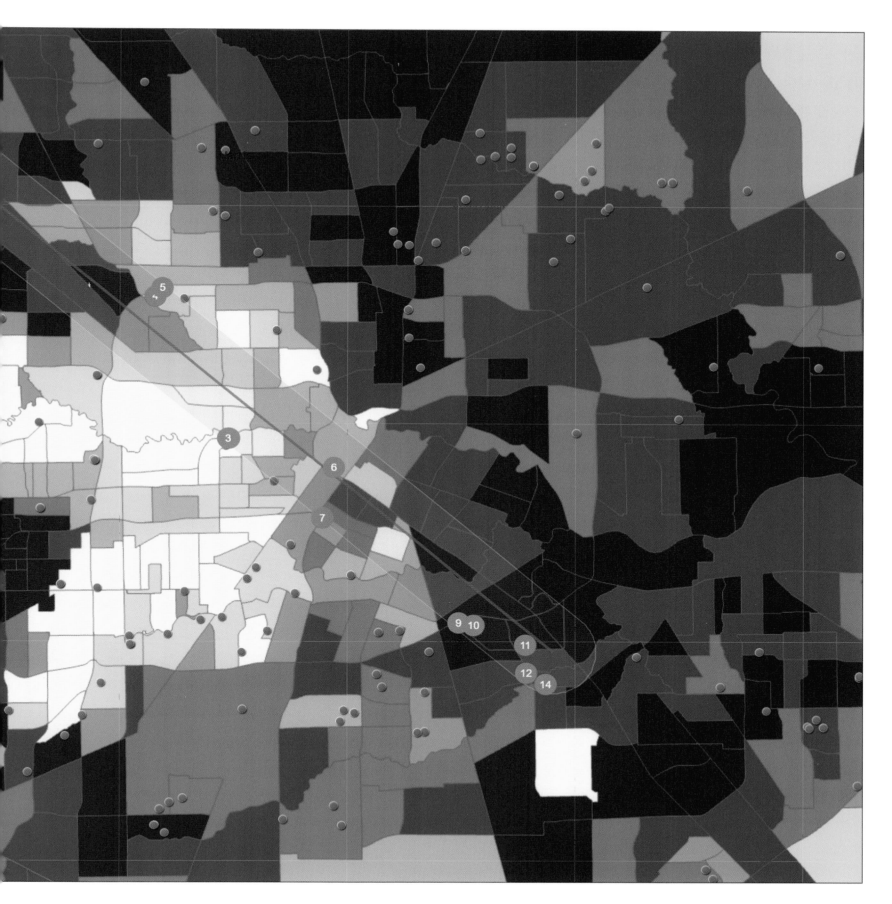

AFTER THE FIRE: WEB APP FOR WILDFIRE RECOVERY

College of Forestry Extension Fire Program, Oregon
State University, Corvallis, Oregon, USA
By Jillian Pihulak, Jenessa Stemke, Jordon Lindsey,
Daniel Leavell, and Carrie Berger

In early September 2020, extreme weather conditions
fueled the explosive growth of wildfires in Oregon.
Growing from 500 acres to over 130,000 acres
overnight, the Beachie Creek Fire was one of five
megafires (fires which exceed 100,000 acres) burning
simultaneously across the state. The Almeda Fire was
only 3,200 acres but burned thousands of homes in just
a few hours. Overall, Oregon's 2020 wildfire season
claimed the lives of 11 people, burned more than a
million acres, and destroyed over 5,000 homes and
non-residential structures.

This map shows the Riverside Fire which burned
138,000 acres and came close to Portland suburbs,
prompting evacuations. The Riverside Fire had 12.3%
high soil burn severity (shown as red in image), which
is above average. Highly impacted soils have a higher
risk of erosion, particularly in the steep western slopes
of Oregon's Cascade Range where annual precipitation
can average up to 75 inches.

To aid in recovery, the College of Forestry Extension
Fire Program at Oregon State University created the
After the Fire StoryMaps story. This online resource
provides information about the impacts of fire and
allows Oregon residents to create maps of their
affected properties to assist in post-fire recovery
planning.

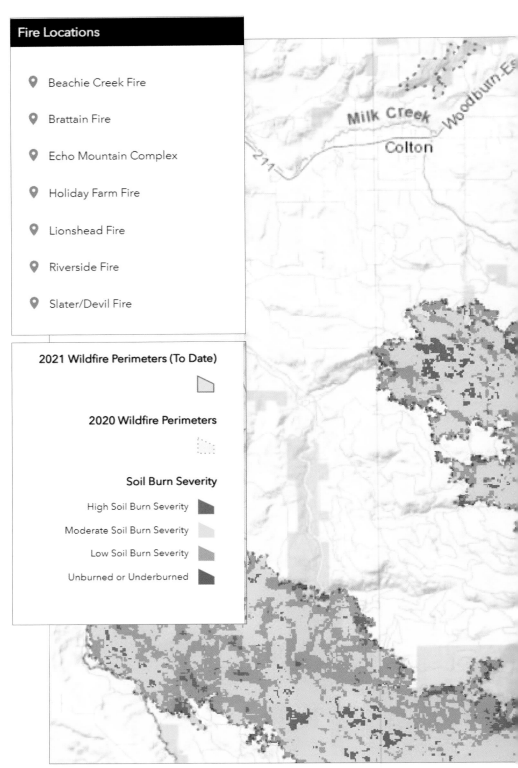

Courtesy of College of Forestry Extension Fire Program, Oregon State University.

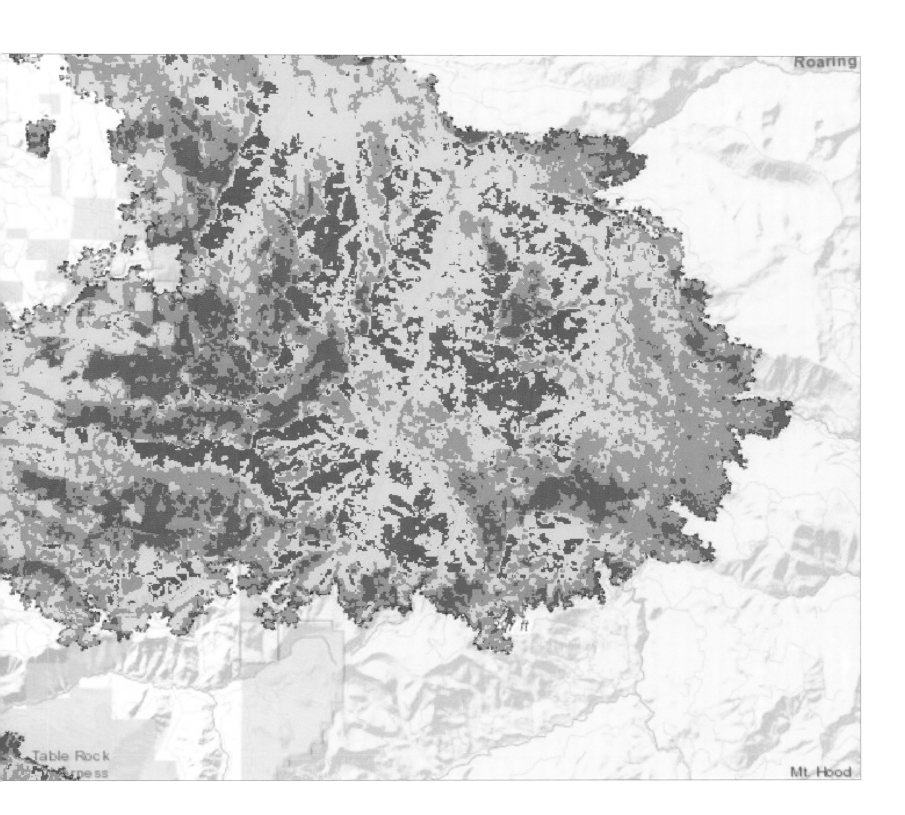

CONTACT
Jenessa Stemke
jenessa.stemke@oregonstate.edu

SOFTWARE
ArcGIS Pro, ArcGIS Online,
ArcGIS StoryMaps

DATA SOURCES
Fire perimeters via Inciweb; Soil Burn Severity via US
Forest Service Burned Area Emergency Response team

COUNTYWIDE 3D BUILDING MODELS

Ayres Associates Inc., Milwaukee, Wisconsin, USA
By Michael Seidel

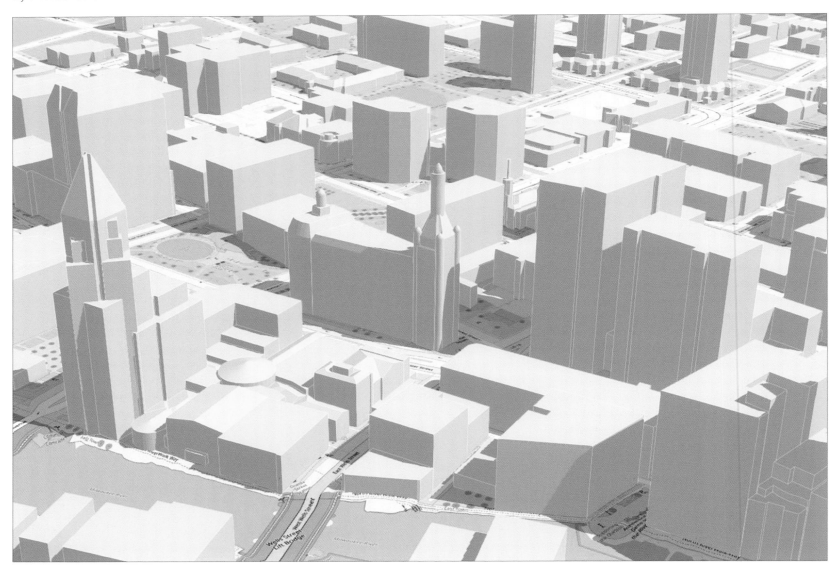

3D building models across the entirety of Milwaukee County, Wisconsin, were derived from 2.2 terabytes of aerial lidar data. The models were created in a semiautomated fashion with special consideration given to buildings of architectural significance and areas of cultural importance. This effort is the first of its kind in the state of Wisconsin.

The final multipatch feature class, more than 400,000 buildings, was published as a scene layer in ArcGIS Online, and an accompanying application was developed to serve the data. This project was an exercise in how to effectively store, process, and visualize vast amounts of point cloud data and represents significant progress in creating useful derivatives from high-density lidar.

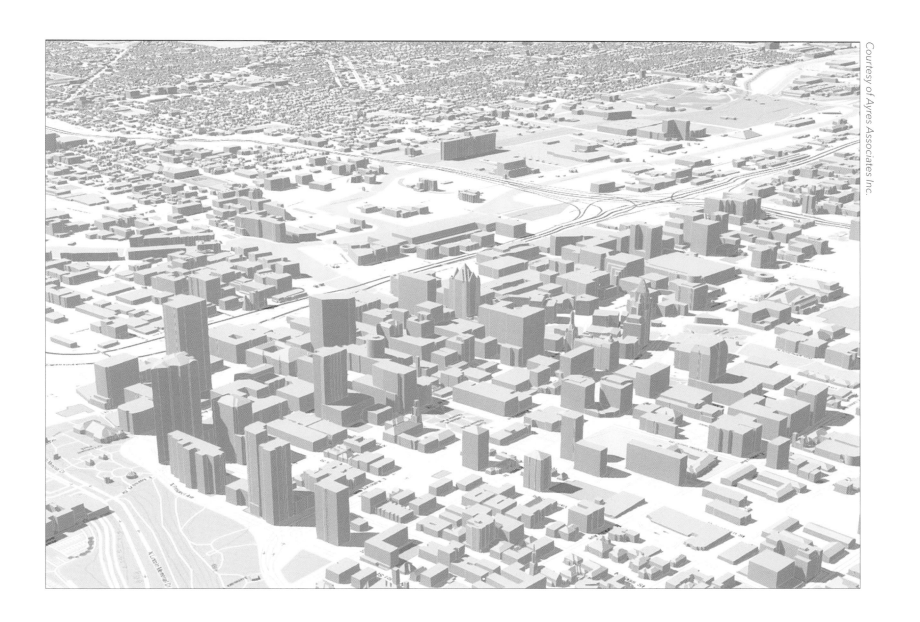

CONTACT
Michael Seidel
seidelm@ayresassociates.com

SOFTWARE
ArcGIS Pro,
ArcGIS 3D Analyst,
ArcGIS Spatial Analyst,
ArcGIS Web AppBuilder,
ArcGIS Online, LASTools

DATA SOURCES
Ayres Associates, Milwaukee Metropolitan
Sewerage District, Southeastern Wisconsin
Regional Planning Commission

MAPPING THE PROPORTION OF PAPUANS AND NON-PAPUANS IN TANAH PAPUA

Indonesian Institute of Sciences
West Papua Province, Indonesia
By Dwiyanti Kusumaningrum

Tanah Papua, or the Land of Papua, is located on the world's largest tropical island and endowed with rich cultural and biological diversity. It consists of the two provinces of Papua and West Papua and is the region with the lowest human development index in Indonesia. In 2019, the Indonesian Institute of Sciences conducted social demographic mapping in Tanah Papua to investigate the provision of health and education services and support the Indonesian government in making inclusive policies for the Papuans.

This map shows the proportion of Papuans and non-Papuans in West Papua Province based on the Indonesian Population Census 2010. The large influx of migrants in urban areas has greatly affected the proportion of native Papuans in their homeland. The number of native Papuans (Orang Asli Papua/OAP) in West Papua Province is 381,933, or 51.45 percent of the total.

CONTACT
Dwiyanti Kusumaningrum
dwiyanti.kusumaningrum@brin.go.id

SOFTWARE
ArcGIS Desktop

DATA SOURCES
The Indonesian Population Census 2010,
Statistics Indonesia

Courtesy of Indonesian Institute of Sciences.

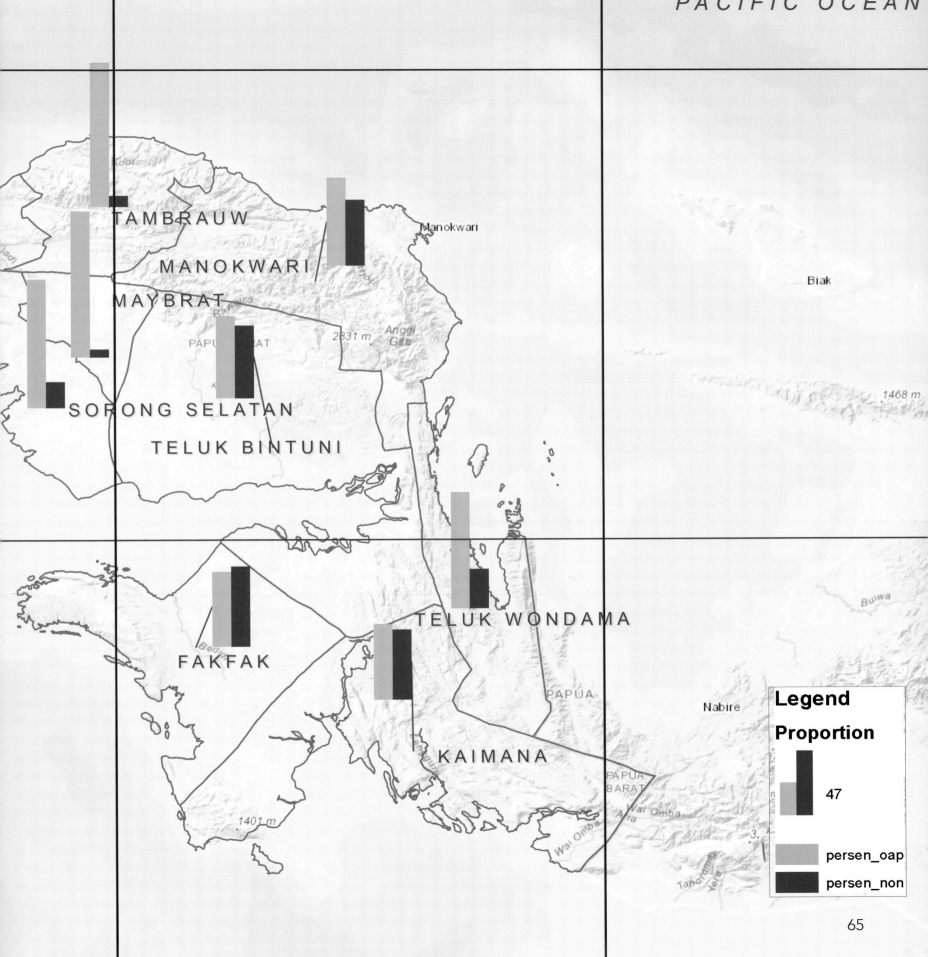

PACIFIC OCEAN

TAMBRAUW

MANOKWARI

MAYBRAT

SORONG SELATAN

TELUK BINTUNI

TELUK WONDAMA

FAKFAK

KAIMANA

Manokwari

Biak

Nabire

Legend

Proportion

47

persen_oap

persen_non

65

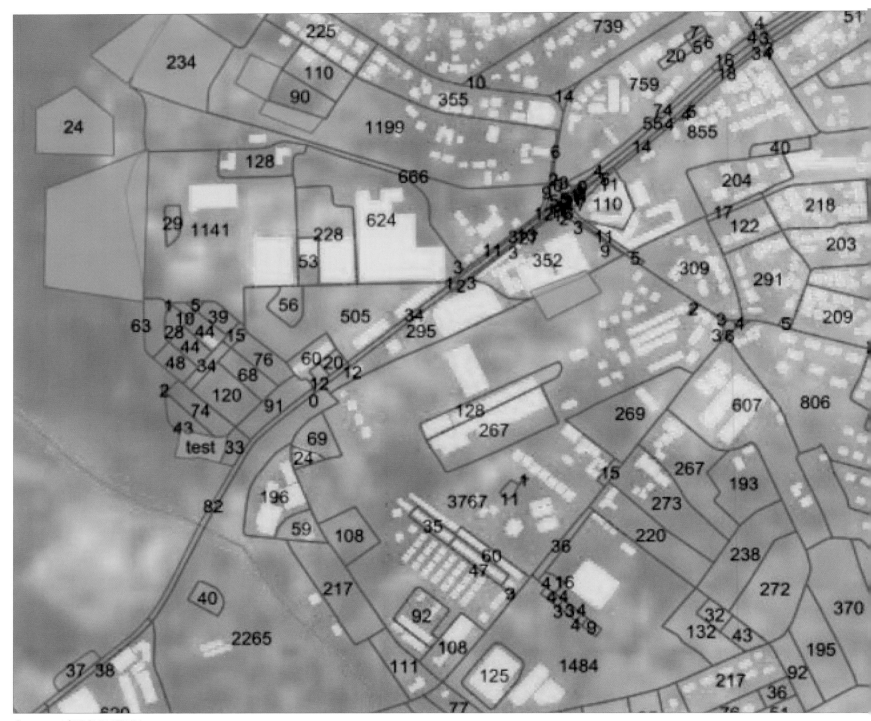

Courtesy of TECHNATIUM.

CADASTRE WITHOUT BORDERS

TECHNATIUM
Massy, France
By Jean Koivogui

Called Cadastre without Borders, this application makes it possible to automatically generate cadastres of cities and municipalities.

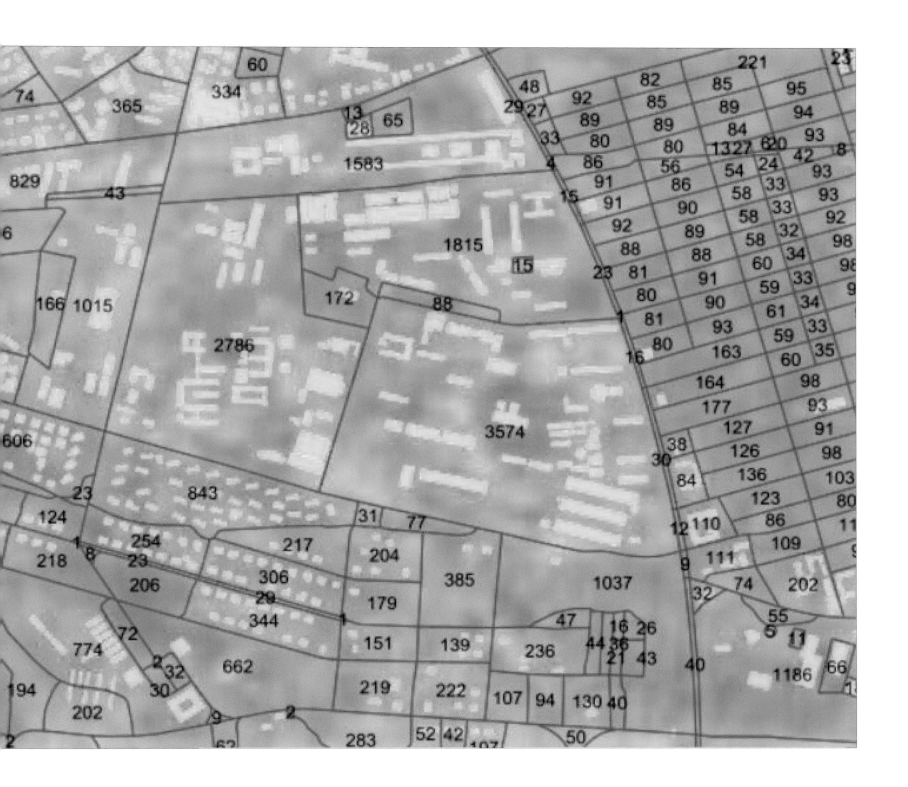

CONTACT
Jean Koivogui
jean.koivogui@nexafrika.com

SOFTWARE
ArcGIS® CityEngine™

DATA SOURCES
Airbus,
Sentinel-type image data

VILNIUS COMMUNITIES: A HOUSING STOCK STRATEGY

Vilnius Tech, Vilnius, Lithuania
By Milda Urbonaviciute, Reda Petraviciute, Polina Kalugina, Ieva Vigelyte, and Justinas Adomaitis

POPULATION INCREASE
DUE TO VILNIUS CITY PLAN

In this age of globalization, major cities are facing multiple issues caused by rapid urbanization. The municipality of Vilnius estimates that 50,000 additional citizens will live in the city by 2050, requiring it to plan for sustainable urban development.

Rapid urbanization has generated specific problems for Vilnius. Some areas of the city have high levels of criminal activity, and other neighborhoods are inhabited by a rapidly aging community. Meanwhile, the central part has an excess of workplaces, and the outskirts lack them. Neighborhoods with the most problematic socioeconomic profiles have become strategic housing development areas.

This study examines the relationship between the city and its citizens. It helps city planners understand the social, economic, and demographic issues and address them with appropriate planning decisions.

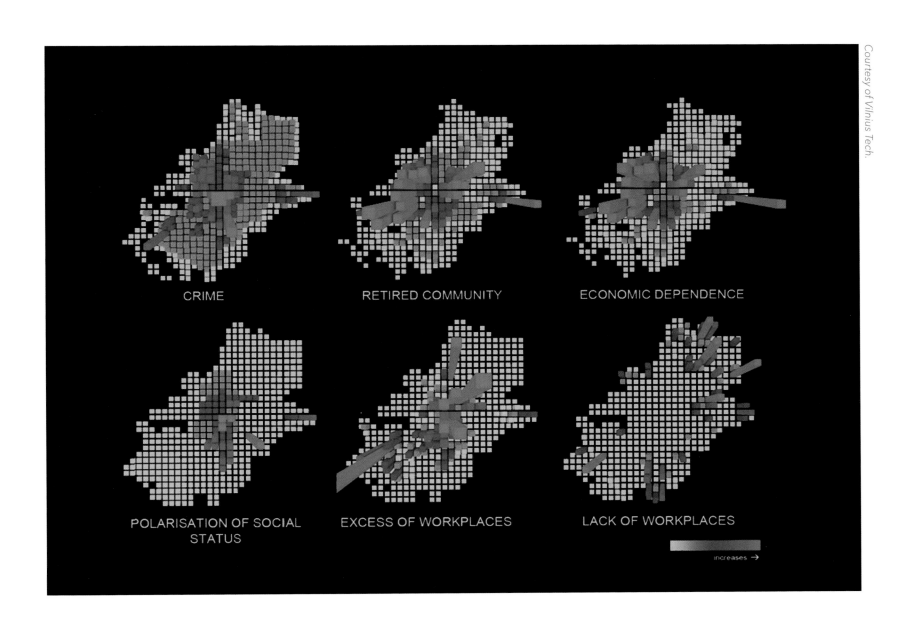

CRIME

RETIRED COMMUNITY

ECONOMIC DEPENDENCE

POLARISATION OF SOCIAL STATUS

EXCESS OF WORKPLACES

LACK OF WORKPLACES

increases →

CONTACT
Milda Urbonaviciute
urbonaviciutemilda@gmail.com

SOFTWARE
ArcGIS Pro

DATA SOURCES
Official Statistics Portal of Lithuania

TREE PLANTING PRIORITIZATION IN ST. LOUIS

City of St. Louis Planning and Urban Design Agency
St. Louis, Missouri, USA
By Jason Whiteley

Trees play an important role in combating urban heat islands by lowering surface temperatures and providing shade to residents. Numerous additional benefits can be associated with tree planting, such as improving community health outcomes, providing carbon sequestration, mitigating climate vulnerability, establishing wildlife corridors, and increasing urban biodiversity.

To guide prioritization of tree-planting efforts, the city's Planning and Urban Design Agency partnered with the city's Office of Sustainability and Forestry Division and developed this Tree Planting Prioritization Suitability Model. It identifies sites that are both suitable for tree planting and have the greatest health equity impact. The suitability model incorporates layers that account for priorities such as existing land cover and trees, the number of emergency room visits because of heat-related illness, residential access to air conditioning, walkable roads with low levels of tree cover, and The Trust for Public Land's Urban Heat Island Index. The final model provides a citywide overview for where to prioritize tree planting, based on these combined factors.

CONTACT
Jason Whiteley
whiteleyj@stlouis-mo.gov

SOFTWARE
ArcGIS Pro, ArcGIS Spatial Analyst

DATA SOURCES
The Trust for Public Land -- Urban Heat Island Index (Summer 2018 and 2019), MoRAP/East-West Gateway Council of Governments – Land Cover (2017), EPA EnviroAtlas – Percent Tree Cover near Walkable Roads (accessed 2020), City of St. Louis Assessor's Office/City of St. Louis Planning and Urban Design Agency – Percent of Buildings without Central AC (2017), City of St. Louis Health Department/ESSENCE – Percentage of ER Visits Due to Heat-Related Conditions (2015–2019)

Courtesy of City of St. Louis Planning and Urban Design Agency.

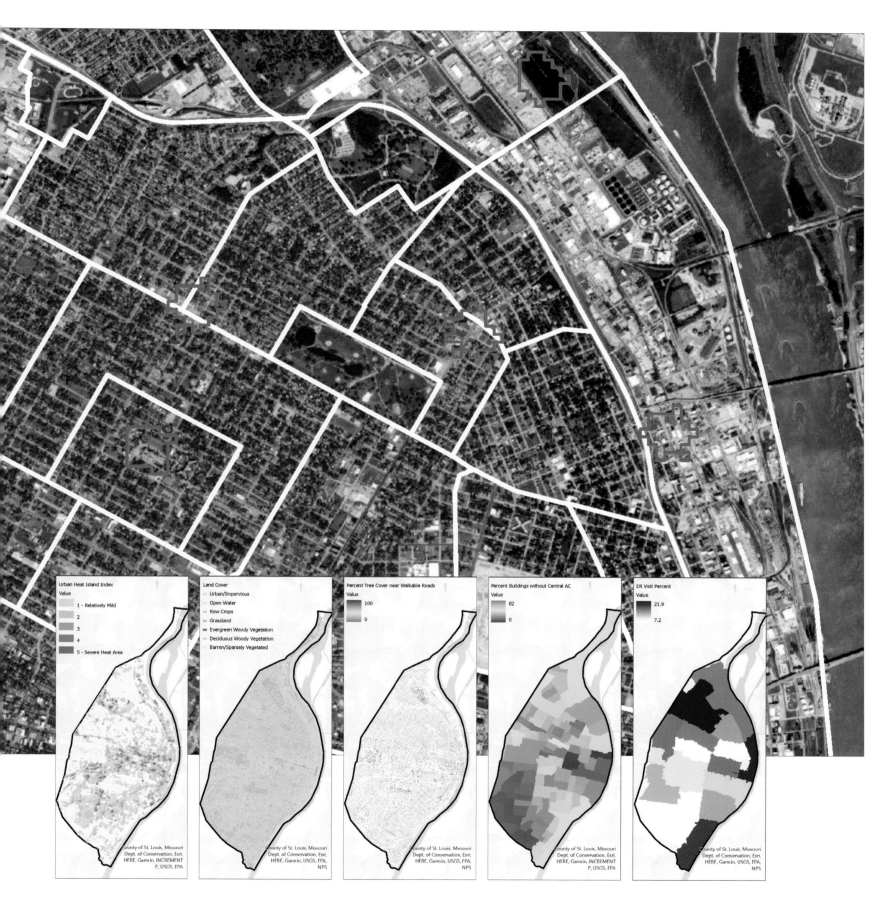

Urban Heat Island Index
Value
1 - Relatively Mild
2
3
4
5 - Severe Heat Area

County of St. Louis, Missouri
Dept. of Conservation, Esri,
HERE, Garmin, INCREMENT
P, USGS, EPA

Land Cover
Urban/Impervious
Open Water
Row Crops
Grassland
Evergreen Woody Vegetation
Deciduous Woody Vegetation
Barren/Sparsely Vegetated

County of St. Louis, Missouri
Dept. of Conservation, Esri,
HERE, Garmin, USGS, EPA,
NPS

Percent Tree Cover near Walkable Roads
Value
100
0

County of St. Louis, Missouri
Dept. of Conservation, Esri,
HERE, Garmin, USGS, EPA,
NPS

Percent Buildings without Central AC
Value
82
0

County of St. Louis, Missouri
Dept. of Conservation, Esri,
HERE, Garmin, USGS, EPA

ER Visit Percent
Value
21.9
7.2

County of St. Louis, Missouri
Dept. of Conservation, Esri,
HERE, Garmin, USGS, EPA,
NPS

THE ART OF PRESIDENTIAL PRIMARY PATTERNS

County of Collin
McKinney, Texas, USA
By Bret Fenster, Husham Mohamed, Vicky Shen, Kendall Holland, Ramona Luster, Tim Nolan, and Abigail Burns

Traditionally, polling locations have been assigned to voters near their precinct, and they have not had the option to vote in other locations throughout the county. Since 2009, "vote centers" have allowed citizens in Collin County the freedom to cast their ballot where it is most convenient for them.

In this map, early voting turnout from the 2020 presidential primary election is shown. Voter distribution is represented with lines going from the voter's address to the vote center where they cast their ballot. Given the option, it appears that people will vote in locations that are not always close to home. This helps election officials project how busy and popular each vote center may be in the future. Popular vote centers will receive more equipment and staffing to meet the projected demand.

Experimental symbology yielded an artistic interpretation of the data on a watercolor basemap. The exported map was enhanced to add an oil painting effect for additional texture. It intentionally contains no labels to reinforce the vision of "data as art."

CONTACT
Bret Fenster
bfenster@co.collin.tx.us

SOFTWARE
ArcGIS Pro, Adobe Photoshop

DATA SOURCES
Collin County Elections Department, Esri

Courtesy of County of Collin government.

UPPSALA MASTER PLAN

Uppsala kommun
Uppsala, Sweden
By Svante Guterstam

Parametric model based on a proposed master plan for the
Sydostra southeastern districts with a summary of floor space
by use.

CONTACT
Svante Guterstam
svante.guterstam@uppsala.se

SOFTWARE
ArcGIS Urban, ArcGIS Pro

DATA SOURCES
Uppsala kommun,
Nivå Landskapsarkitektur

Courtesy of Uppsala kommun.

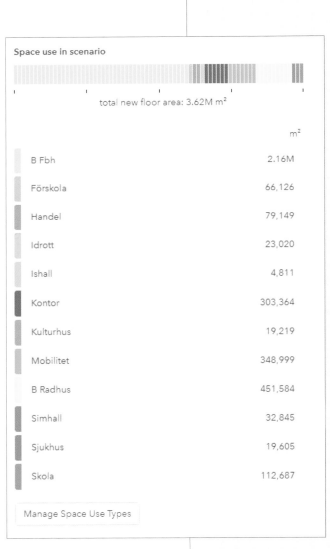

Space use in scenario

total new floor area: 3.62M m²

	m²
B Fbh	2.16M
Förskola	66,126
Handel	79,149
Idrott	23,020
Ishall	4,811
Kontor	303,364
Kulturhus	19,219
Mobilitet	348,999
B Radhus	451,584
Simhall	32,845
Sjukhus	19,605
Skola	112,687

Manage Space Use Types

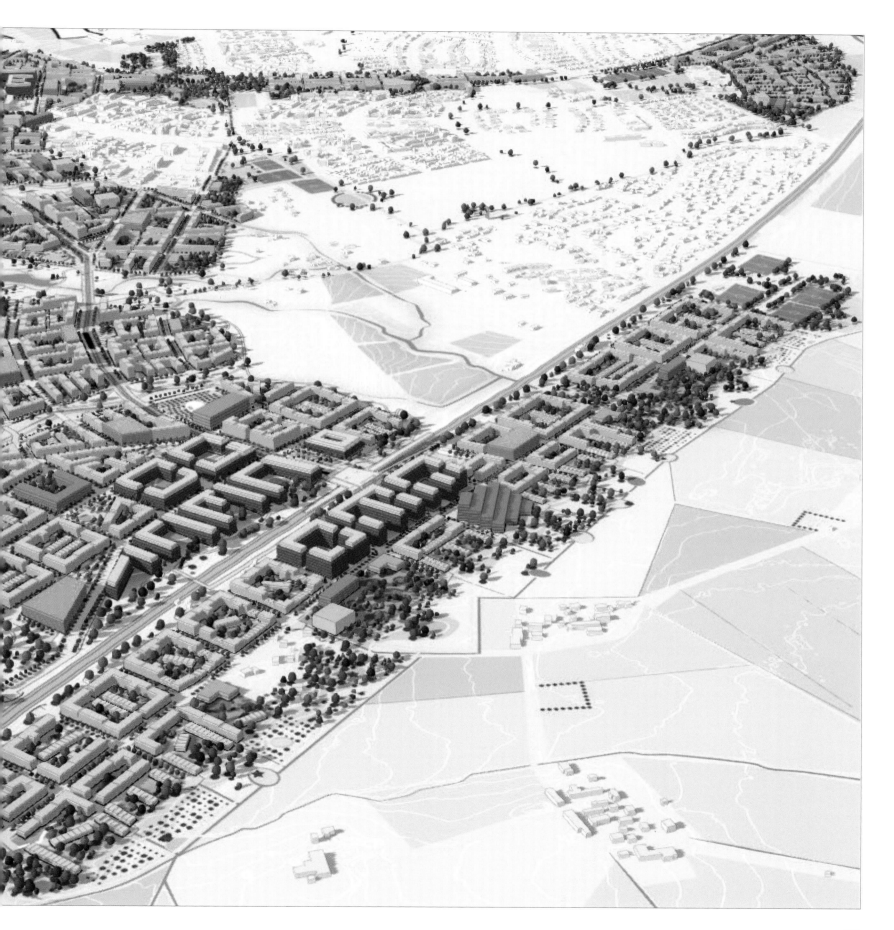

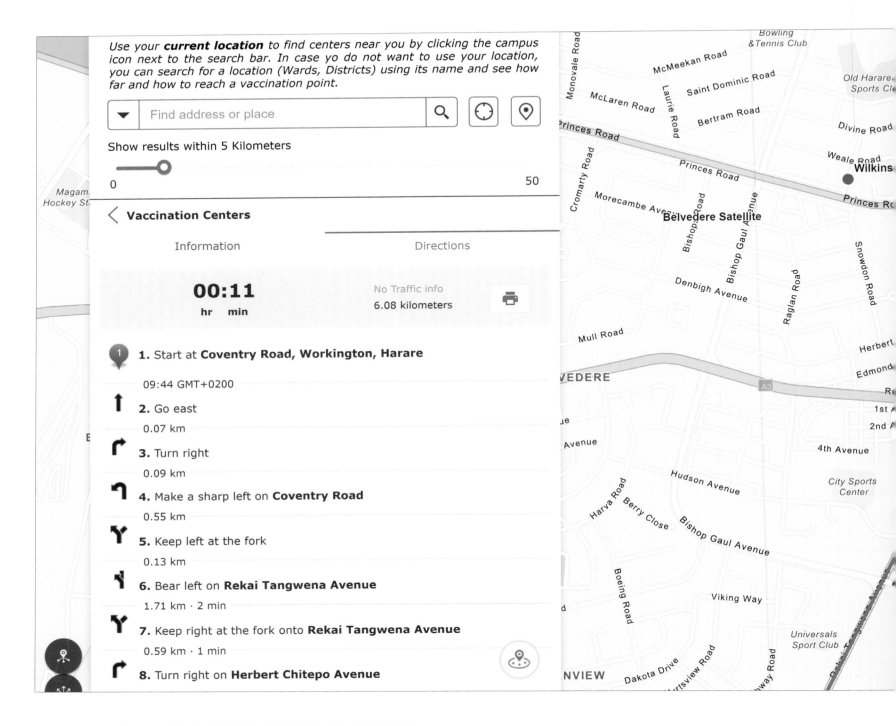

FINDING A VACCINE IN ZIMBABWE

African Surveyors Connect
Central Business District, Harare,
Zimbabwe
By Kumbirai Matingo

With the introduction of coronavirus (COVID-19) vaccines around the world, the people of Zimbabwe have begun a vaccination campaign to achieve herd immunity. Because of financial constraints, not all health centers can provide COVID-19 vaccines. Some provide only one COVID-19 vaccine, whereas others provide all available vaccines. Citizens need this

information to plan their visits and find the closest centers where they can get the vaccine without wasting too many resources or time.

The map shows a snapshot from the COVID-19 Hub for Zimbabwe initiative, with a user receiving location information about the nearest facility to get

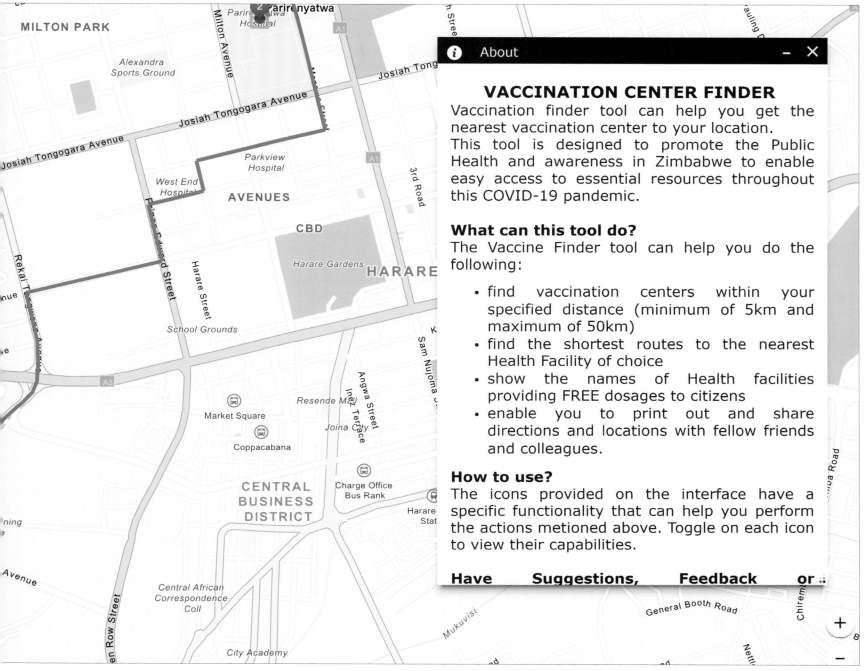

VACCINATION CENTER FINDER

Vaccination finder tool can help you get the nearest vaccination center to your location.

This tool is designed to promote the Public Health and awareness in Zimbabwe to enable easy access to essential resources throughout this COVID-19 pandemic.

What can this tool do?

The Vaccine Finder tool can help you do the following:

- find vaccination centers within your specified distance (minimum of 5km and maximum of 50km)
- find the shortest routes to the nearest Health Facility of choice
- show the names of Health facilities providing FREE dosages to citizens
- enable you to print out and share directions and locations with fellow friends and colleagues.

How to use?

The icons provided on the interface have a specific functionality that can help you perform the actions metioned above. Toggle on each icon to view their capabilities.

Have Suggestions, Feedback or ...

Courtesy of African Surveyors Connect.

a COVID-19 vaccine. Users can view all information about the nearest facility and be given the best route to get there. Because all resources are readily available with this type of location application, hesitancy and slow vaccine uptake are reduced.

CONTACT
Kumbirai Matingo
matingonk@gmail.com

SOFTWARE
ArcGIS Developer,
ArcGIS Enterprise,
ArcGIS Online,
ArcGIS Web AppBuilder

DATA SOURCES
Ministry of Health and Child Care, Johns Hopkins University

UNABLE TO DELIVER: ACCESS TO OBSTETRIC CARE IN NORTHERN MINNESOTA

Carleton College
Northfield, Minnesota, USA
By Ada Wright and Sean MacDonell

Between 2000 and 2015, 15 rural Minnesota hospitals closed their obstetric wards and stopped obstetric services for expectant mothers. This spatial analysis of access to obstetric wards in northern Minnesota (Lake, Cook, and St. Louis Counties) uses a modified gravity model to address the needs of these mothers and their access to obstetric health care. Access scores were developed using the number of ob-gyns at the nearest hospital, travel time to the nearest obstetric ward, and population of women of childbearing age. This analysis was done to identify areas that need increased obstetric outreach as low access to maternal health care has been linked to higher child and mother mortality.

CONTACT
Ada Wright
wrighta2@carleton.edu

SOFTWARE
ArcGIS Desktop, ArcGIS Pro

DATA SOURCES
National Rural Health Association, American Journal of Obstetrics and Gynecology, The Journal of the American Board of Family Practice, The Journal of the American Board of Family Medicine, MPR News, The Canadian Geographer/Le Geographe Canadien, Progress in Spatial Analysis.

Courtesy of Carleton College.

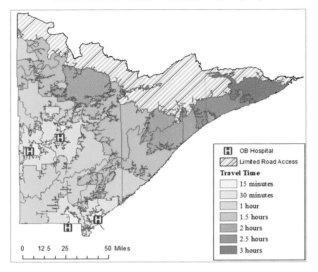

Travel Time to Nearest Obstetric Care (OB)

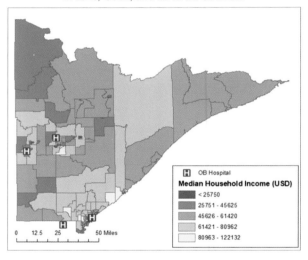

Medina Household Income by Census Block Group in Lake, Cook, and St. Louis Counties

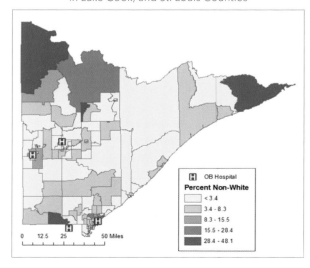

Percent of Non-White Population by Census Block Group in Lake Cook, and St. Louis Counties

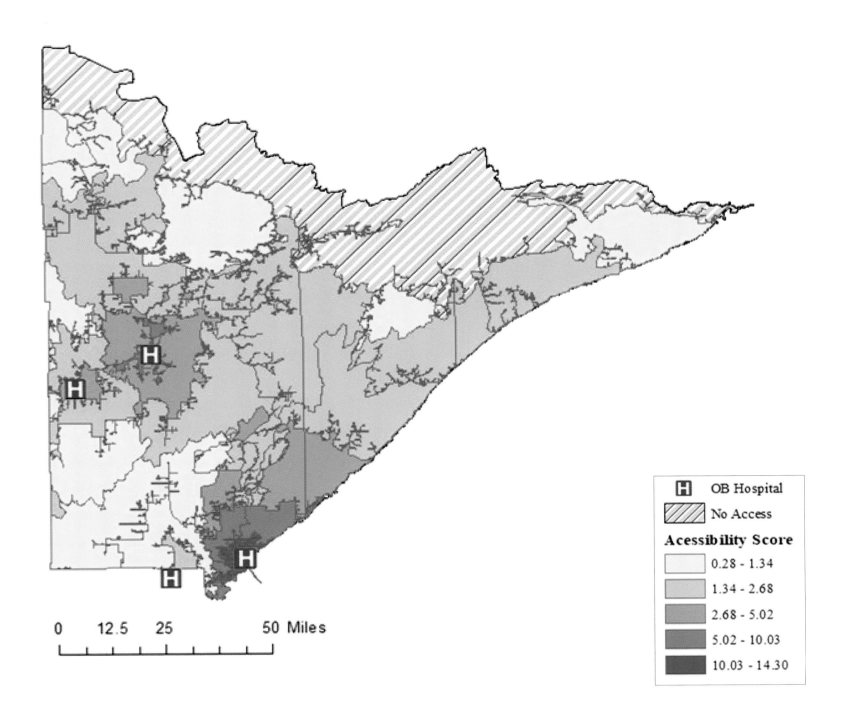

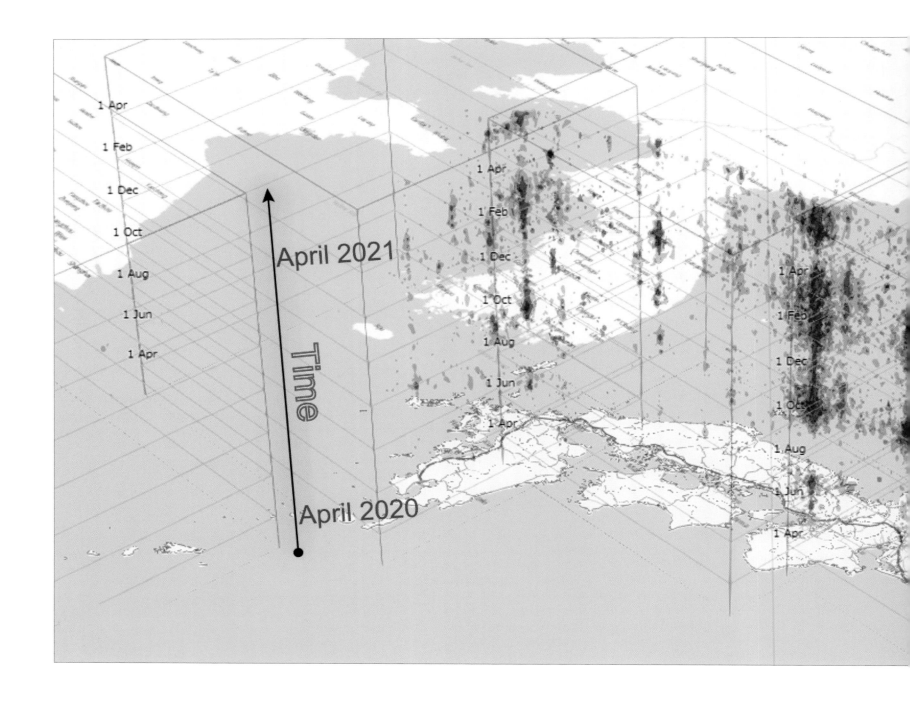

WHERE HAVE JAPANESE COVID-19 OUTBREAKS BEEN SUSTAINED?

Tohoku University
Sendai, Japan
By Tomoki Nakaya
and Shohei Nagata

This 3D interactive map simultaneously visualizes both the geographic and temporal aspects of the COVID-19 outbreak in Japan. The areas in red, blue, and gray correspond to high, medium, and low densities of infection, respectively. Points closest to the basemap occurred at the

beginning of the study period and continue in sequential slices. The densest column indicates a high incidence of infection and suggests that the chain of infection within the area is continuing.

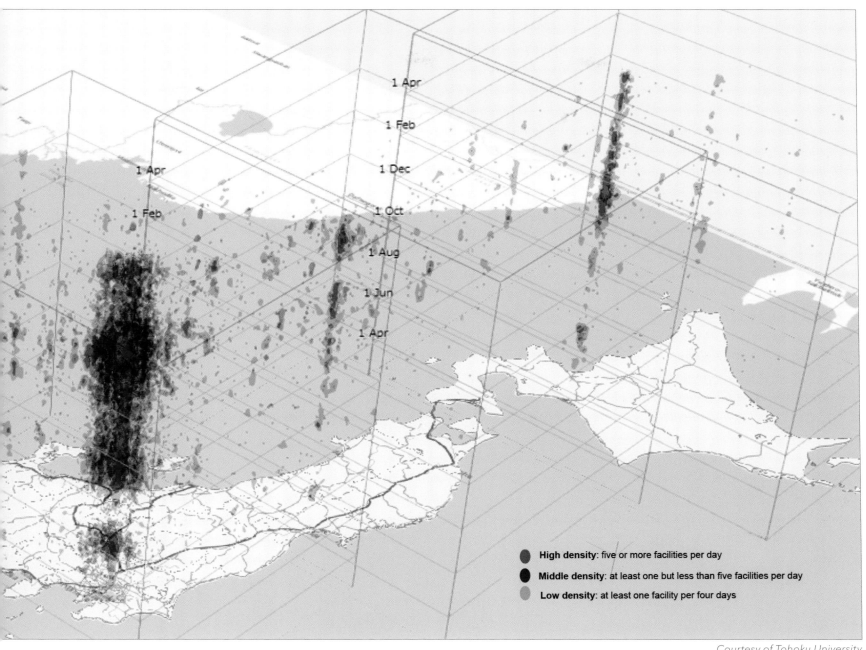

1 Apr
1 Feb
1 Dec
1 Oct
1 Aug
1 Jun
1 Apr

1 Apr
1 Feb

High density: five or more facilities per day

Middle density: at least one but less than five facilities per day

Low density: at least one facility per four days

Courtesy of Tohoku University.

CONTACT

Tomoki Nakaya

tomoki.nakaya.c8@tohoku.ac.jp

SOFTWARE

ArcGIS Pro, ArcGIS 3D Analyst,
ArcGIS API for Python,
ArcGIS API for JavaScript, ArcGIS Online,
Space-Time Density Tool for ArcGIS Pro
 (developed by Tomoki Nakaya for
Tohoku University)

DATA SOURCES

Data assembled by JX Press Corporation
from information on the occurrence of
infections spontaneously posted on the
internet by various institutions

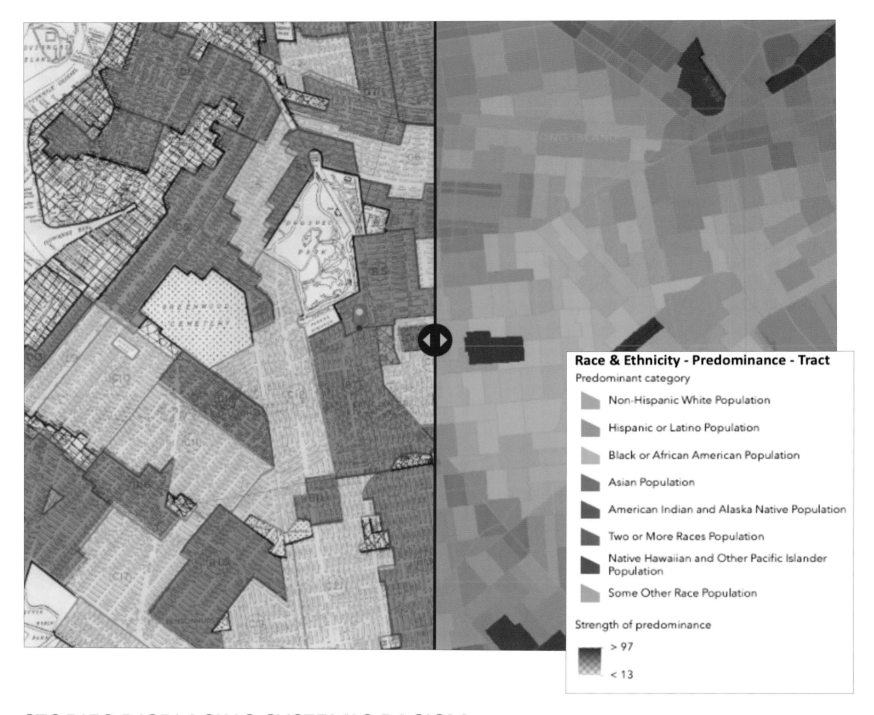

Race & Ethnicity - Predominance - Tract

Predominant category

- Non-Hispanic White Population
- Hispanic or Latino Population
- Black or African American Population
- Asian Population
- American Indian and Alaska Native Population
- Two or More Races Population
- Native Hawaiian and Other Pacific Islander Population
- Some Other Race Population

Strength of predominance

- > 97
- < 13

STORIES DISPLACING SYSTEMIC RACISM

The FaithX Project
Germantown, Maryland, USA
By Ken Howard

"Redlining" was outlawed by the Fair Housing Act in 1968, but the racial segregation it caused persists more than 50 years later. The FaithX Project uses interactive storytelling maps to address pernicious social problems, such as systemic racism. Systemic racism is built on a false yet deeply embedded and nearly unconscious social narrative, based on a framing of "us against them." It must be displaced by consciously embedding a different story in the social narrative—a "we and us" story that is more powerful, infectious, and enduring.

Why does "story" matter? Research shows that data alone does not change hearts, minds, or behaviors

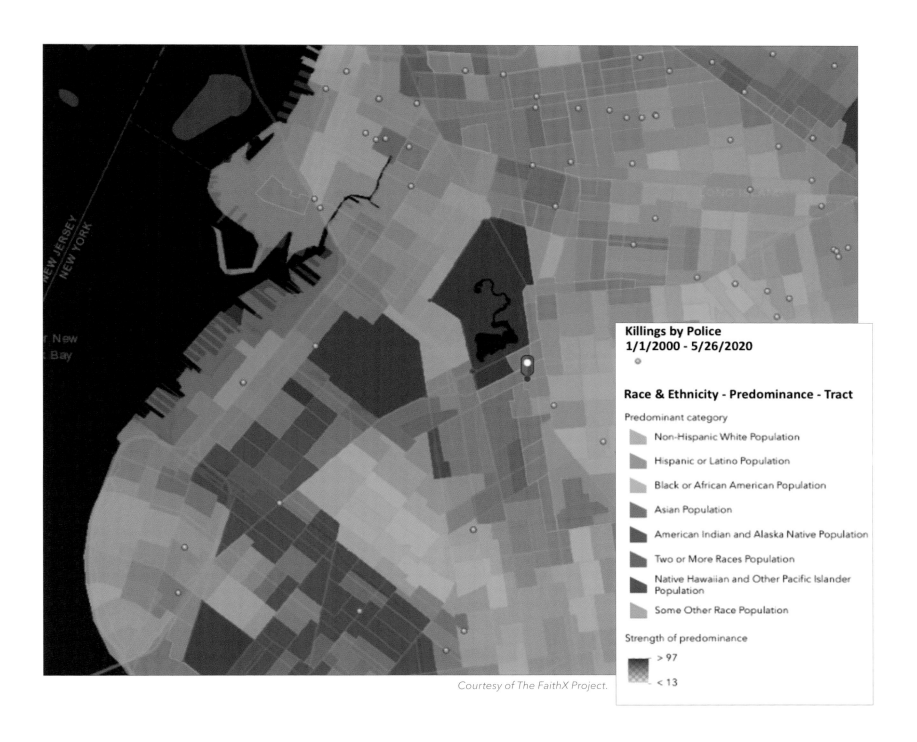

Killings by Police
1/1/2000 - 5/26/2020

Race & Ethnicity - Predominance - Tract

Predominant category

Non-Hispanic White Population

Hispanic or Latino Population

Black or African American Population

Asian Population

American Indian and Alaska Native Population

Two or More Races Population

Native Hawaiian and Other Pacific Islander Population

Some Other Race Population

Strength of predominance

> 97

< 13

Courtesy of The FaithX Project.

unless it tells a story, Stories bring meaning to data. By creating an opportunity for people to interact with mapped data that shows the impacts of systemic racism in their own neighborhoods, it enables them to view the story of systemic racism and their place in it, moving the issue from head to heart, and empowers them to address it.

CONTACT
Ken Howard
ken@faithx.net

SOFTWARE
ArcGIS Pro, ArcGIS
StoryMaps, MapDash

DATA SOURCES
Esri, Historical Home
Owners Loan Corporation

THE POTENTIAL EXTENT OF ANTHRAX IN THE RUSSIAN ARCTIC UNDER A FUTURE CLIMATE SCENARIO

Federal Centre for Animal Health (Vladimir, Russia) and Lomonosov Moscow State University (Moscow, Russia),
By Fedor Korennoy

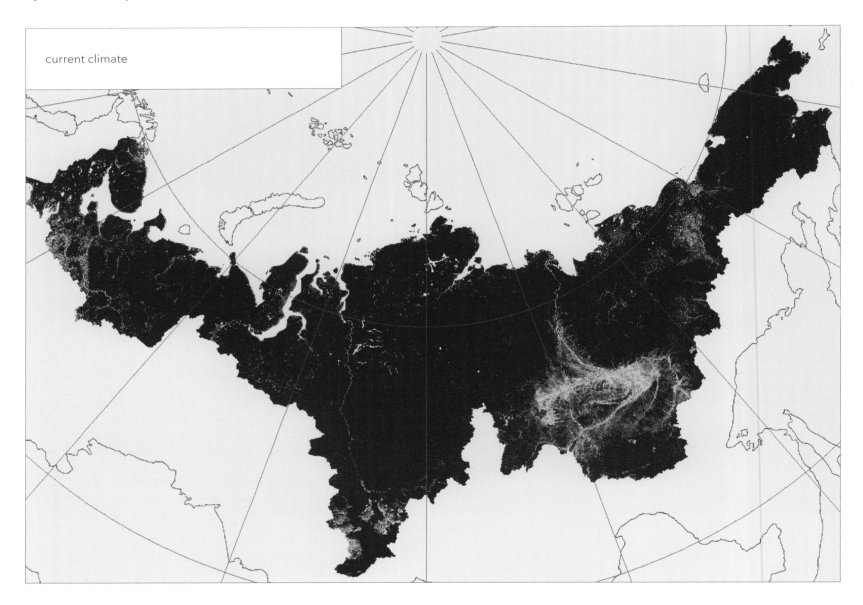

current climate

This map shows locations in the Russian Arctic where anthrax could reemerge based on a projected climate where there are no significant reductions in emissions until 2100. This study assumes that locations where animals have contracted anthrax (present in the soil) and been buried may be treated as indicators of favorable environmental conditions for anthrax. Because of expected warming and thawing of permafrost, these and other sites may become potential sources of infection. Areas with a similar set of environmental parameters may be categorized as potentially dangerous.

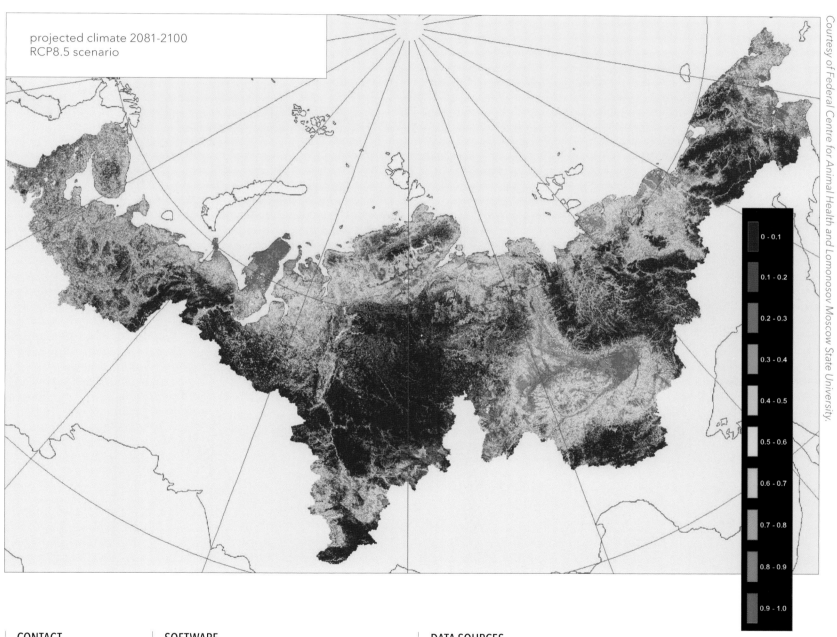

projected climate 2081-2100
RCP8.5 scenario

0 - 0.1
0.1 - 0.2
0.2 - 0.3
0.3 - 0.4
0.4 - 0.5
0.5 - 0.6
0.6 - 0.7
0.7 - 0.8
0.8 - 0.9
0.9 - 1.0

CONTACT
Fedor Korennoy
korennoy@arriah.ru

SOFTWARE
ArcGIS Desktop,
ArcGIS Desktop extensions,
MaxEnt (maximum entropy
Environmental Niche Modeling software)

DATA SOURCES
Official register of stationary anthrax
burials in the Russian Federation, Russian
national register of soils, weather station
records, CMIP-5 project

TRAVEL TIME TO HEALTH FACILITIES IN ZAMBIA

Columbia University
New York City, New York, USA
By Matthew Heaton

An example from a national-scale mapping exercise, this map focuses on Gwembe, a district in the Southern province of Zambia. The map was created on behalf of the COVAX COVID-19 immunization campaign and features population, settlement, and health facility layers.

This travel time map shows which settlements have difficulty accessing health services. By simulating the time required to traverse a 100-by-100-meter area based on elevation change, land cover, and other input parameters, it shows the time it would take a resident of each settlement to travel to their nearest health facility via mixed-mode transportation. Population counts are disaggregated by age and distance in minutes to the nearest health facility. Health facility service areas were created using a composite of travel time and settlement population.

The full 245-page atlas was designed for COVID-19 vaccination microplanning. However, because the maps are being distributed on poster-size vinyl meant to last, they can be repurposed for routine immunization planning, general navigation, or other uses. With these maps as a resource, health workers can more effectively target and allocate resources for the most vulnerable and remote populations and ensure more equitable distribution of care.

CONTACT
Matthew Heaton
mheaton@ciesin.columbia.edu

SOFTWARE
ArcGIS Pro

DATA SOURCES
Fraym, GRID3 Data Hub, 2018 Zambia Demographic and Health Survey base survey

Courtesy of Columbia University.

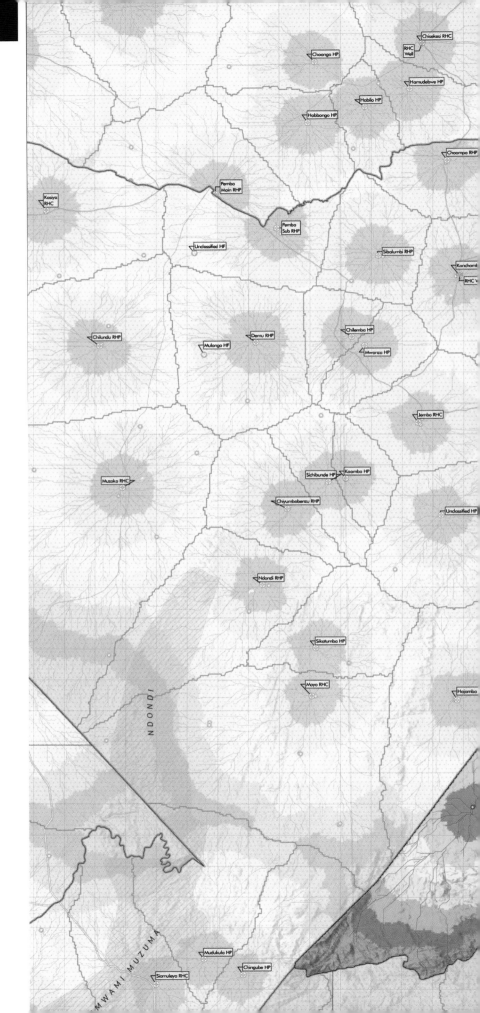

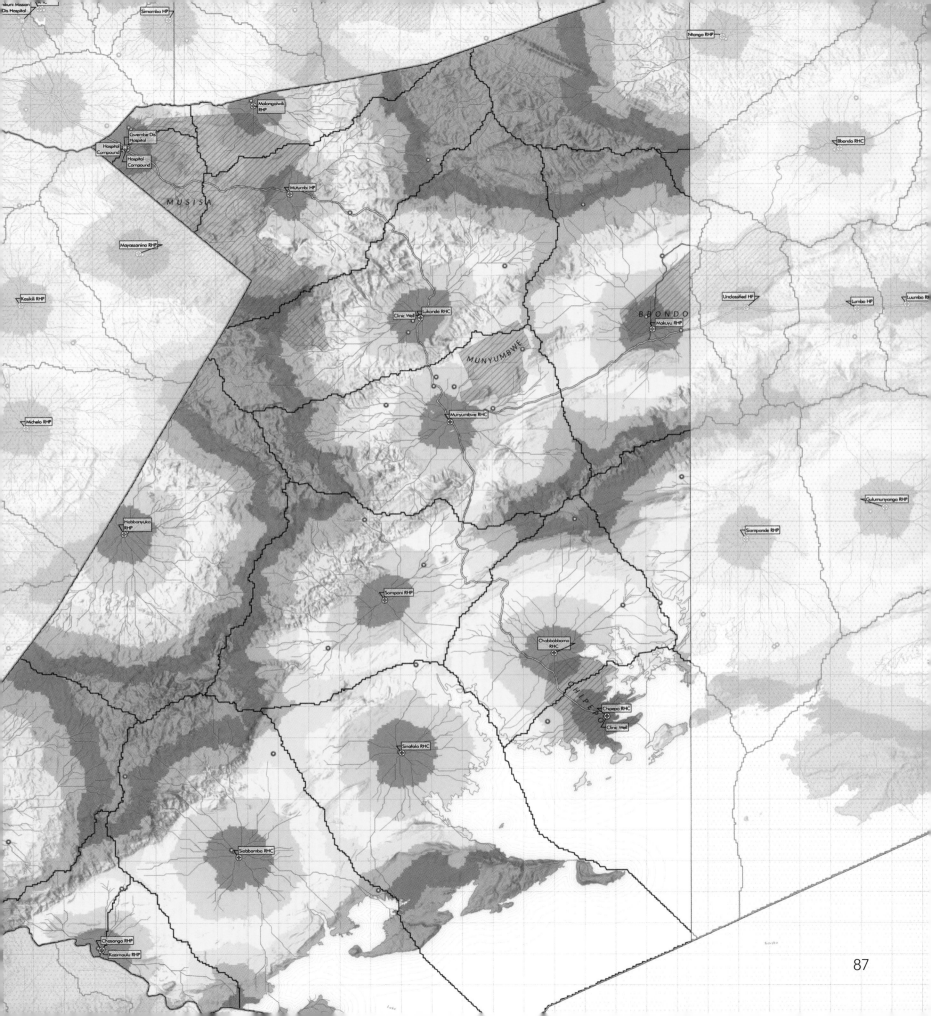

CASTING A BALLOT ON THE NAVAJO NATION

San Diego State University
San Diego, California, USA
By Nicolas Vadun-Lemp

Mail-in voting, a process often touted as a solution to low turnout rates, especially during the COVID-19 pandemic, can instead create more barriers for voting on reservations. Because of their remoteness, low density, and lack of paved roads, reservations rarely have at-home mail delivery. This means a vote by mail is a vote by post office, which can be a significant distance away.

This map attempts to visualize ballot access for the Navajo (Diné) Nation. The drive time to post offices is shown along with blue points indicating home locations. A picture emerges of how many people face a significant journey to vote.

CONTACT

Nicolas Vadun-Lemp
nickvlemp@gmail.com

SOFTWARE

ArcGIS Pro, ArcGIS® Network Analyst™

DATA SOURCES

US Census Bureau, US Post Office

Courtesy of San Diego State University.

10 People

Post Offices

Minutes by Car to P.O.

0-10

10-20

20-30

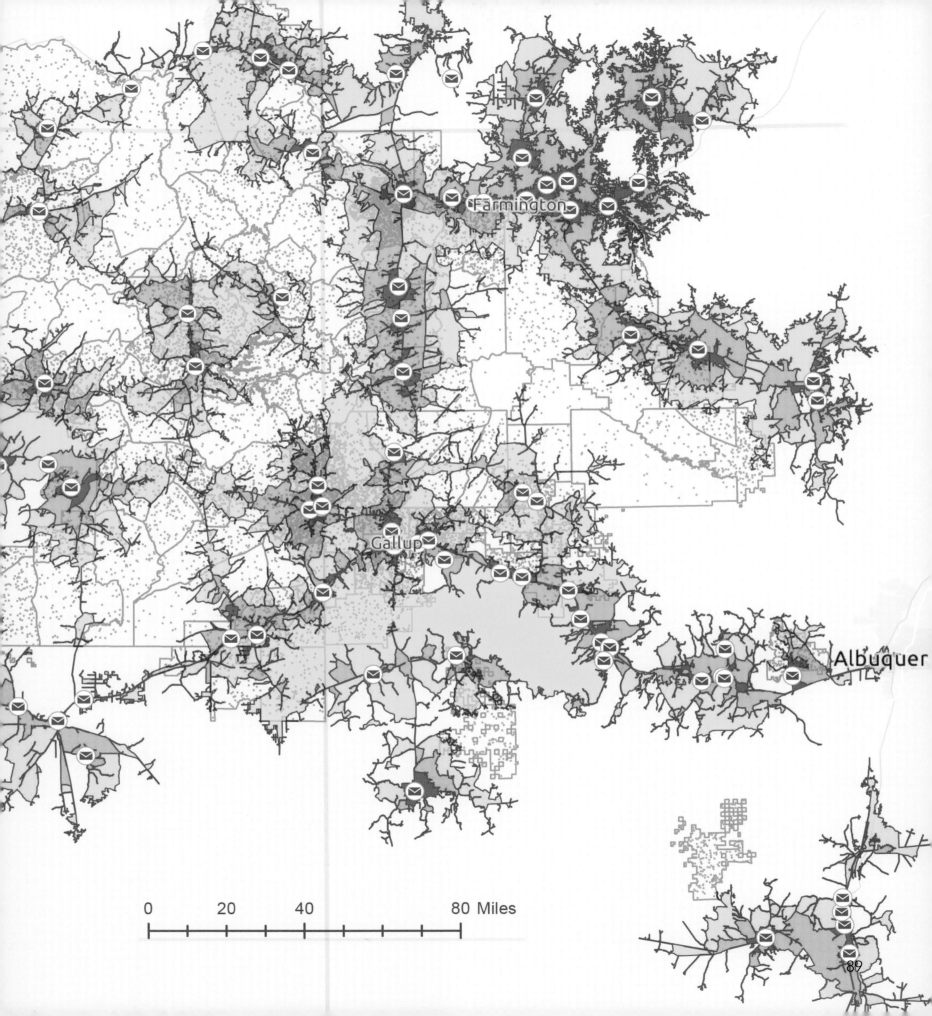

Farmington

Gallup

Albuquer

0   20   40   80 Miles

89

CORN CROP LOCATIONS IN MATO GROSSO, BRAZIL

SeerAI Inc.
Aurora, Colorado, USA
By Daniel Wilson

This map is an output of an artificial intelligence/machine learning model identifying locations of the second-season corn crop safrinha using spatiotemporal patterns in satellite imagery. The map shows the likely locations of safrinha corn across the Brazilian state of Mato Grosso and can be used to estimate the total viable area of corn.

CONTACT
Daniel Wilson
dwilson@seerai.space

SOFTWARE
ArcGIS Online, ArcGIS StoryMaps, ArcGIS Pro, SeerAI Geodesic Platform

DATA SOURCES
MODIS-MCD43A4

Courtesy of SeerAI Inc.

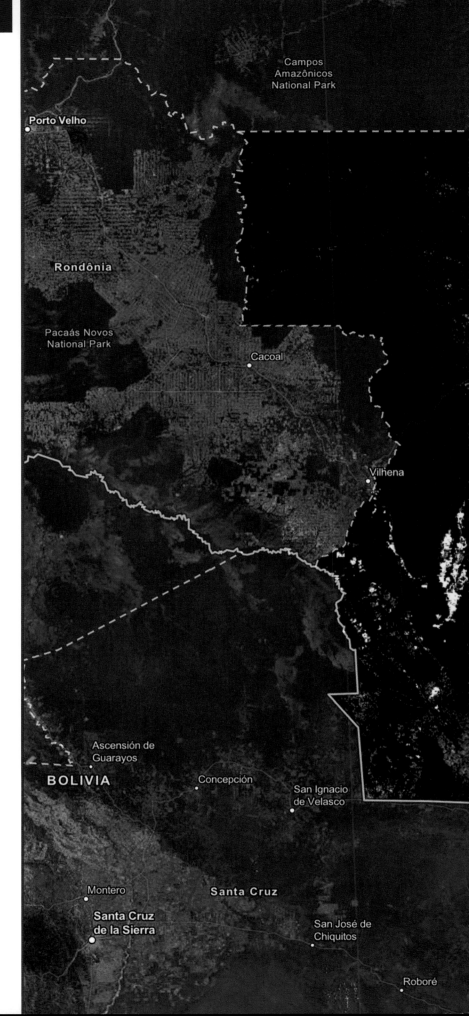

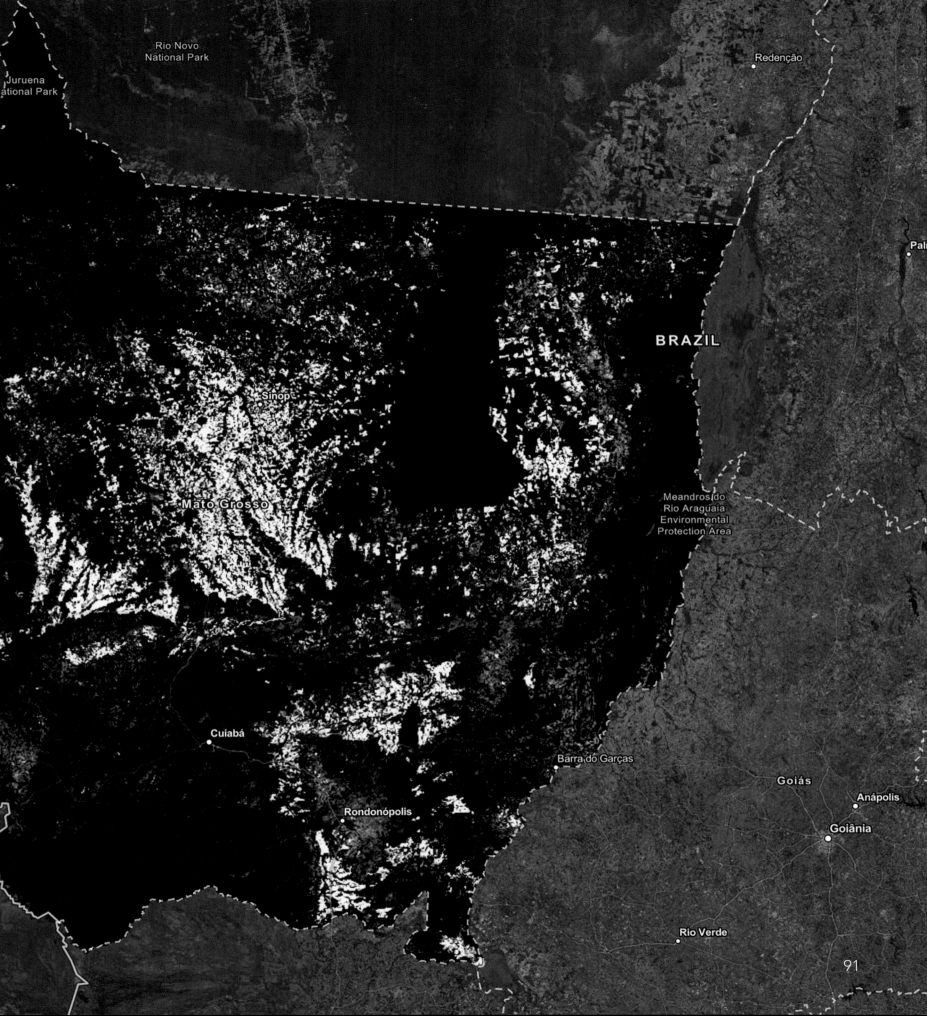

FARMING LOCATIONS FOR THE NAVAJO NATION

Elijah Allan
Flagstaff, Arizona, USA
By Elijah Allan

The ancestral homelands of the Navajo (Diné) Nation are dominated by high desert, semi-arid lands, which have been progressively impacted by severe drought events and changes in the climate.

The Navajo Nation has a limited food economy, and the food system currently in place is not adequate to support it as an absolute sovereign nation. The 2012 Census of Agriculture indicated there were 5,247 citizen-owned and operated cropland farms on the Navajo Nation, comprising 154,119 acres. Navajo Agriculture Products Industry indicates they have around 70,000 acres of irrigated farmland, with a goal of 110,630 acres.

The goal of this project was to do a raster analysis of Land Cover, Precipitation, Elevation, Slope, and Soil-Available Water Storage to identify farming suitability and help guide and improve agricultural and food economy decisions.

CONTACT
Elijah Allan
allanel@oregonstate.edu

SOFTWARE
ArcGIS Desktop, ArcGIS Spatial Analyst

DATA SOURCES
Navajo Nation Land Department, Oregon State University PRISM, USGS GAP, USDA NRCS

Courtesy of Elijah Allan.

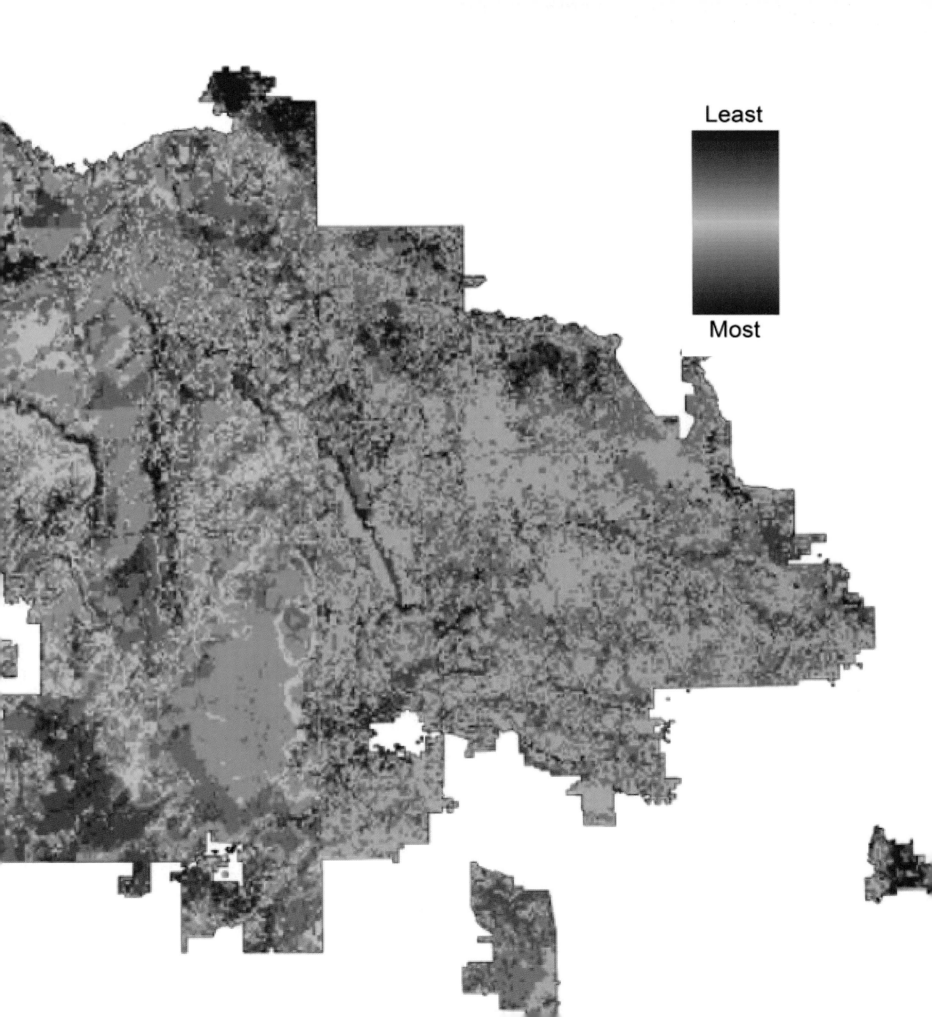

Least

Most

MAIN EXPORT ROUTES FOR BRAZILIAN SOYBEANS

USDA Foreign Agricultural Service
Washington, DC, USA
By Ryan Williams

Brazil is one of the most significant producers in the world oilseed market. Created by contract staff under the International Production Assessment Division of the USDA for the Agricultural Marketing Service's Brazil Soybean Transportation Guide, this map shows the soybean export routes from various regions of Brazil and highlights the expansion of soybean production and infrastructure within the Amazon ecoregion.

This map—along with other information on transportation, economics, production, exports, and infrastructure development provided in the Soybean Transportation Guide: Brazil 2020—helps guide important decisions and prompt new research.

CONTACT
Ryan Williams
ryant.williams@usda.gov

SOFTWARE
ArcGIS Pro, Microsoft Office

DATA SOURCES
World Wildlife Fund, Brazilian Institute of Geography and Statistics, USDA Agricultural Marketing Service and Foreign Agricultural Service

Courtesy of USDA Foreign Agricultural Service.

➡ **Export route**

Ⓐ **Rail terminal**

Ⓢ **Port**

— **Major road**

— **Major river**

▢ **Amazon ecoregion** [1]

Soybean production dens

Metric tons per square kilomet

< 30
31 - 80
81 - 145
146 - 215
> 216

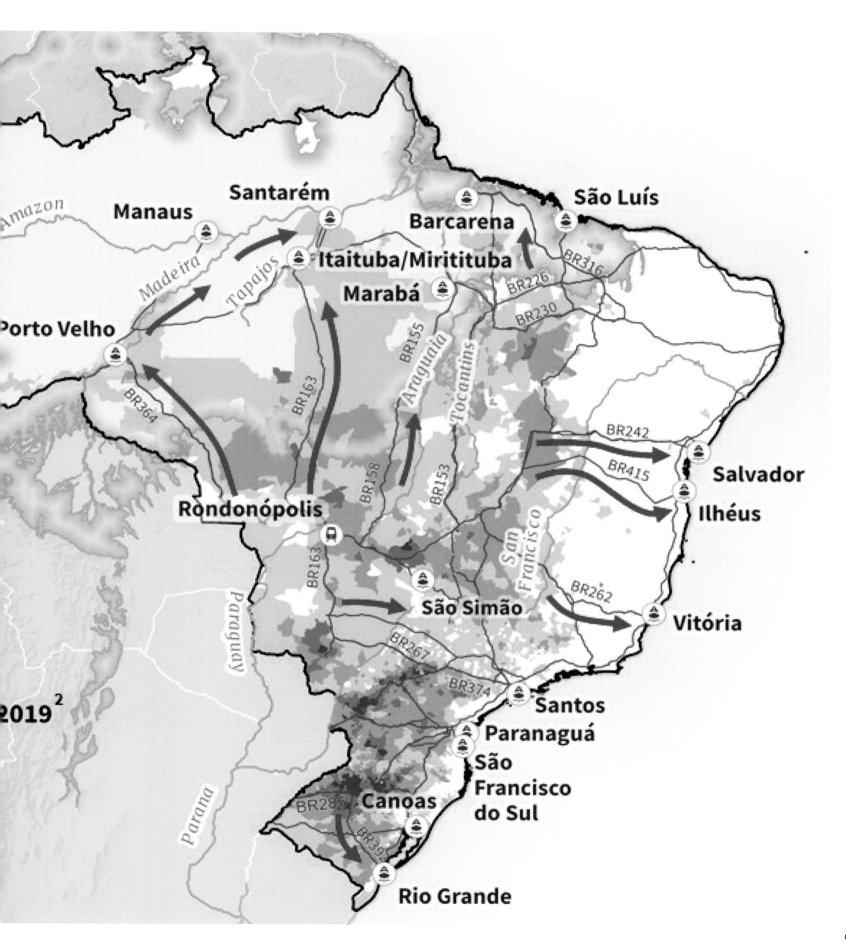

CARTOGRAPHIC EXPLORATIONS OF NON-EARTH PLANETARY SURFACES

Planetary Modeling Inc., Kirkland, Washington, USA
By Eian Ray

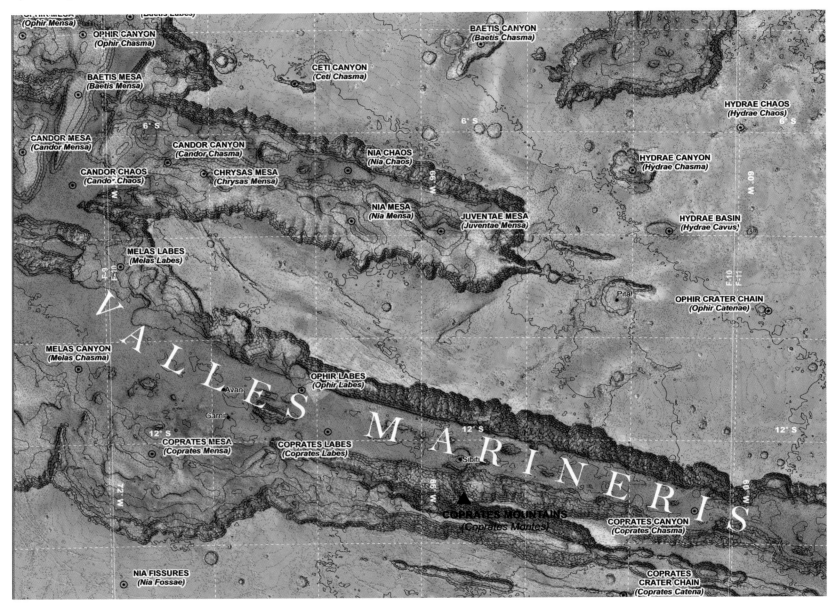

Planetary surfaces like those of Mars and Earth's moon are a source of infinite inspiration for scientists, students, and casual observers seeking to reimagine contemporary human exploration. The featured maps show a portion of Valles Marineris, an immense rift valley on the Martian equator, and Mare Crisium, one of many vast lunar seas. These maps represent two of a 495-map plate collection of Mars, Earth's moon, and Mercury, modeled after traditional Earth atlases.

Although the maps portray expansive alien landscapes, they also show many familiar features. For example, mountains, valleys, and canyons are all visible and recognizable in their respective environments and are brought to life here. The imagery accentuates topographical features, displays a combination of true and false color, and aids in the visual understanding of geological and environmental processes. As access to data from planetary surfaces increases, modern mapping technology will continue to advance a "boots on the ground" experience.

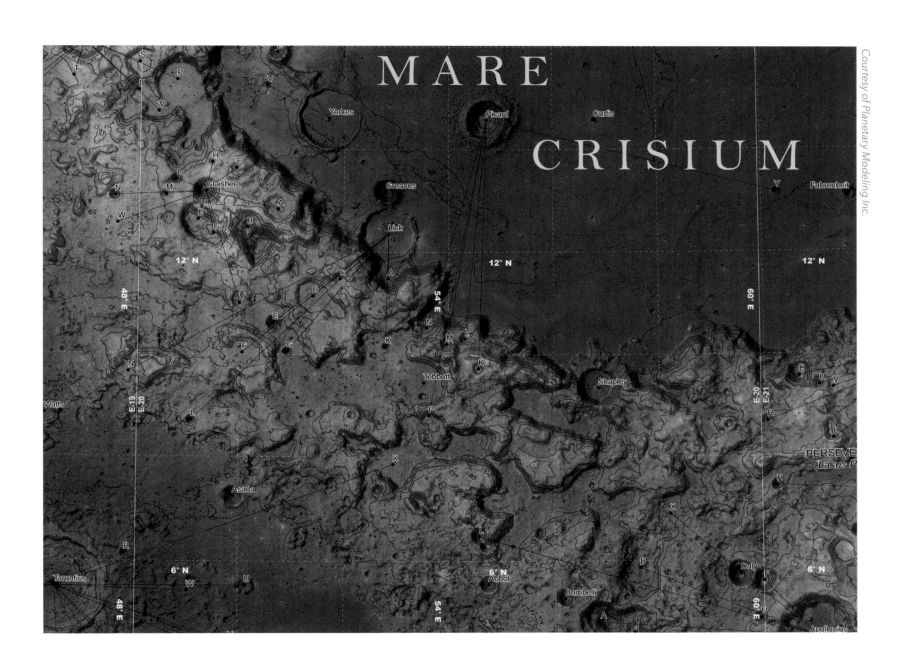

CONTACT
Eian Ray
info@redmapper.com

SOFTWARE
ArcGIS Desktop

DATA SOURCES
Mars Orbital Laser Altimeter Digital Elevation Model, Mars Viking Colorized Global Mosaic Satellite Imagery, Lunar Orbital Laser Altimeter digital elevation model, Lunar Clementine UVVISv2 Hybrid Mosaic satellite imagery, International Astronomical Union Planetary Nomenclature Database, Human Activity and Imprint Geodatabase

REMAPPING THE ELEVATION OF THE UNITED STATES

NOAA, US Department of Commerce
Boulder, Colorado, USA
By Brian Shaw

NOAA's National Geodetic Survey is currently working on modernizing the National Spatial Reference System (NSRS). This effort will replace the current US datums, including the North American Datum of 1983 (NAD 83), the North American Vertical Datum of 1988 (NAVD 88), and other vertical datums. This modernization will correct elevations (orthometric heights) and horizontal positions by more than a meter in some places. This map shows the approximate estimated changes in elevation for the Conterminous United States (CONUS) between NAVD 88 and the North American–Pacific Geopotential Datum of 2022 (NAPGD2022).

CONTACT
Brian Shaw
brian.shaw@noaa.gov

SOFTWARE
ArcGIS Desktop, Adobe Photoshop

DATA SOURCES
Esri, NOAA, USGS

Courtesy of USDOC NOAA.

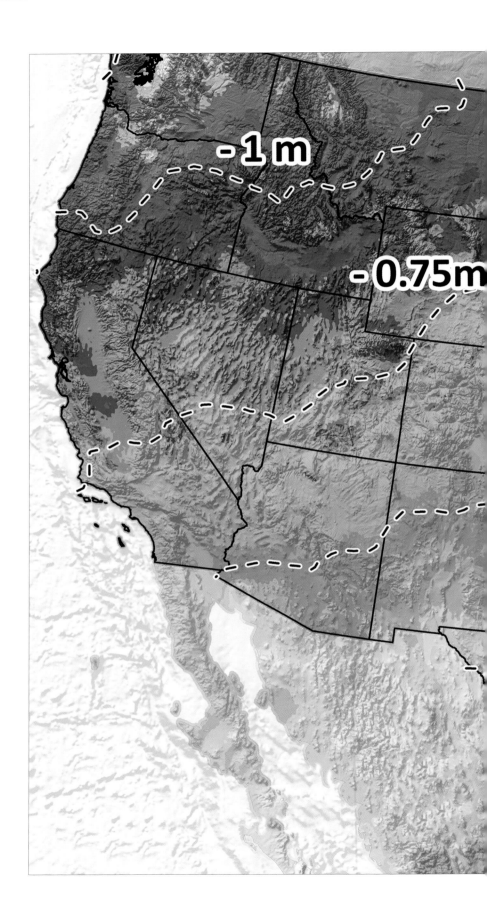

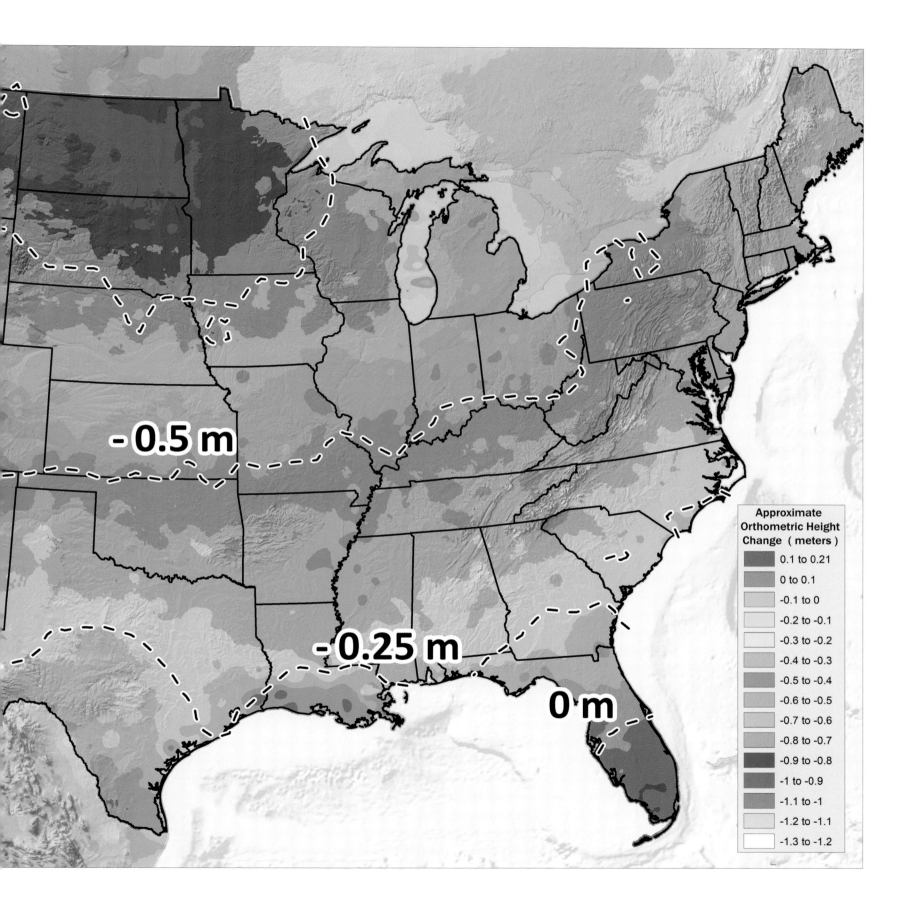

GEOLOGIC MAP OF THE MCDOWELL MOUNTAINS, MARICOPA COUNTY, ARIZONA

McDowell Sonoran Conservancy
Scottsdale, Arizona, USA
By Steven J. Skotnicki

Dr. Steve Skotnicki developed the first comprehensive geologic map of the McDowell Mountains in Scottsdale, Arizona, by hand drawing during his field observations over almost 10 years. Volunteers of the McDowell Sonoran Conservancy, working under the guidance of Skotnicki and geologist Brian Gootee of the Arizona Geological Survey, digitized this extremely complex map, including colorizing and labeling it and adding the detailed map unit descriptions. Skotnicki's work refined and corrected the existing, much older preliminary surveys of the area and documented previously unknown features. The map includes roughly 3,000 individual map units and 86 rock types. The resulting map was published by the Arizona Geological Survey and is available electronically.

CONTACT
Dan Gruber
dgruber@alum.mit.edu

SOFTWARE
ArcGIS Desktop, ArcGIS Desktop extensions, specialized ArcGIS Desktop toolbars for geologic mapping symbology developed by the Arizona Geological Survey and USGS

DATA SOURCES
Dr. Steve Skotnicki, USGS, City of Scottsdale, Maricopa County.

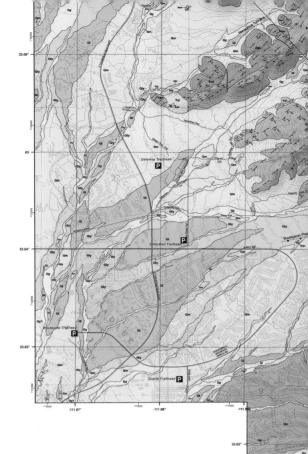

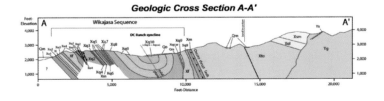

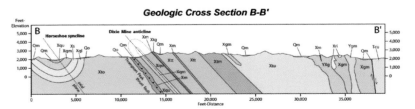

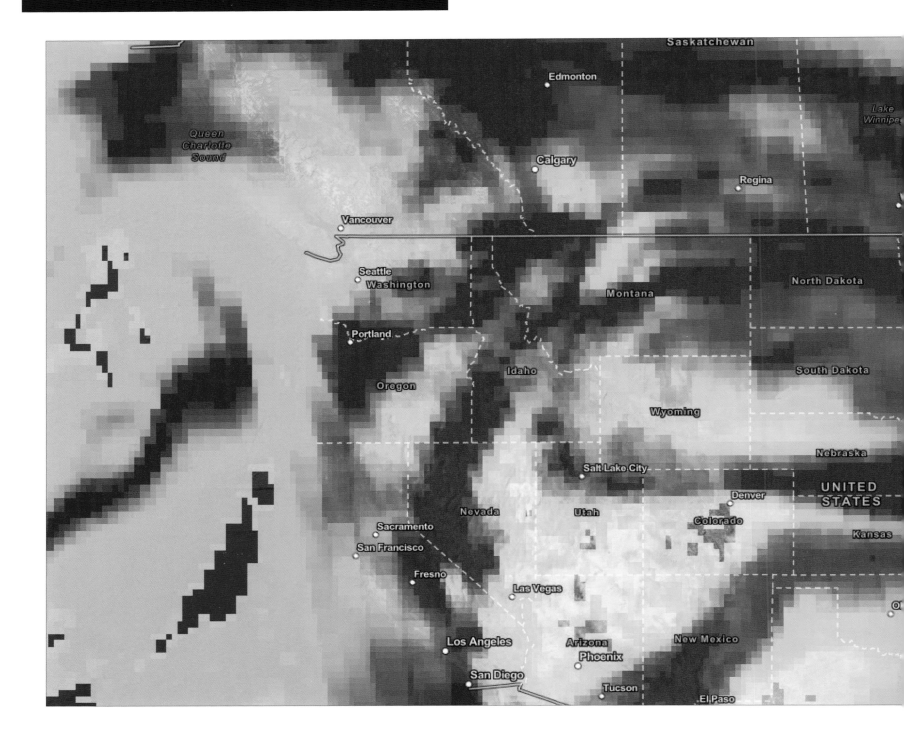

MAPPING SMOKE IMPACT FROM WESTERN US FIRES

NASA
Silver Spring, Maryland, USA
By Garrett Layne

The 2020 western United States fire season was one of the worst on record, particularly in California which experienced its largest wildfire season in modern history. The impacts of the fire season were felt across the entire continent as smoke affected air quality and visibility on the East Coast of the United States.

The NASA Earth Applied Sciences Disasters Program used the Ozone Mapping and Profiler Suite (OMPS) instrument on the Suomi National Polar-orbiting Partnership (Suomi NPP) satellite to monitor how the smoke moved across the continent.

This map shows the OMPS Aerosol Index on September 13, 2020. The aerosol index indicates the

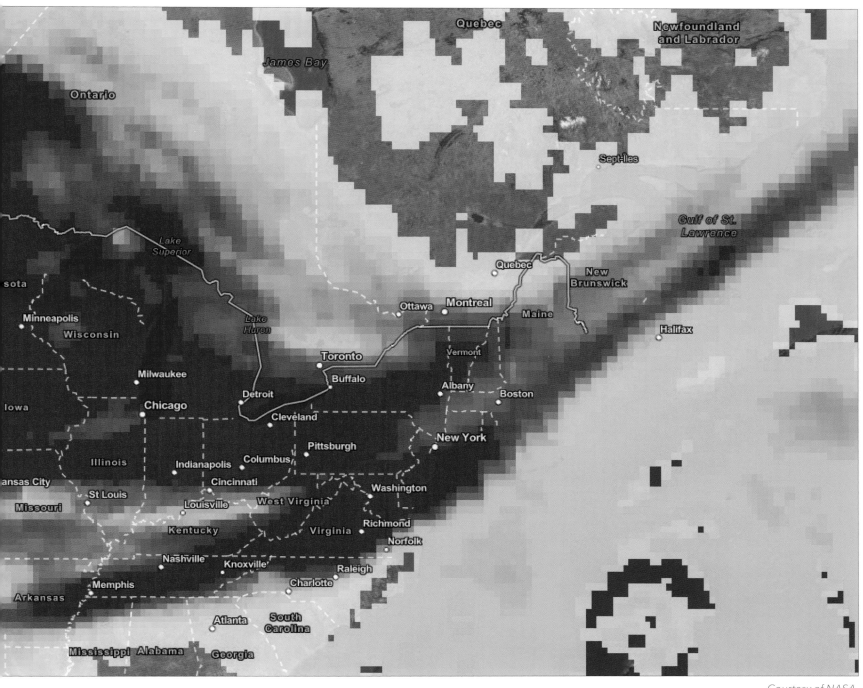

Courtesy of NASA .

presence of ultra violet (UV)-absorbing particles in the air with areas in dark red indicating high concentrations of aerosols that can reduce visibility or impact human health. This snapshot is from a time-enabled web application that monitors the evolution and transport of the smoke plumes as they travel across the continent.

CONTACT
Garrett Layne
garrett.w.layne@nasa.gov

SOFTWARE
ArcGIS Configurable Apps, ArcGIS Desktop, ArcGIS Enterprise, ArcGIS Enterprise portal, ArcGIS Image Server, ArcGIS Online, ArcGIS Pro, ArcGIS StoryMaps, ArcGIS Web AppBuilder

DATA SOURCES
NASA

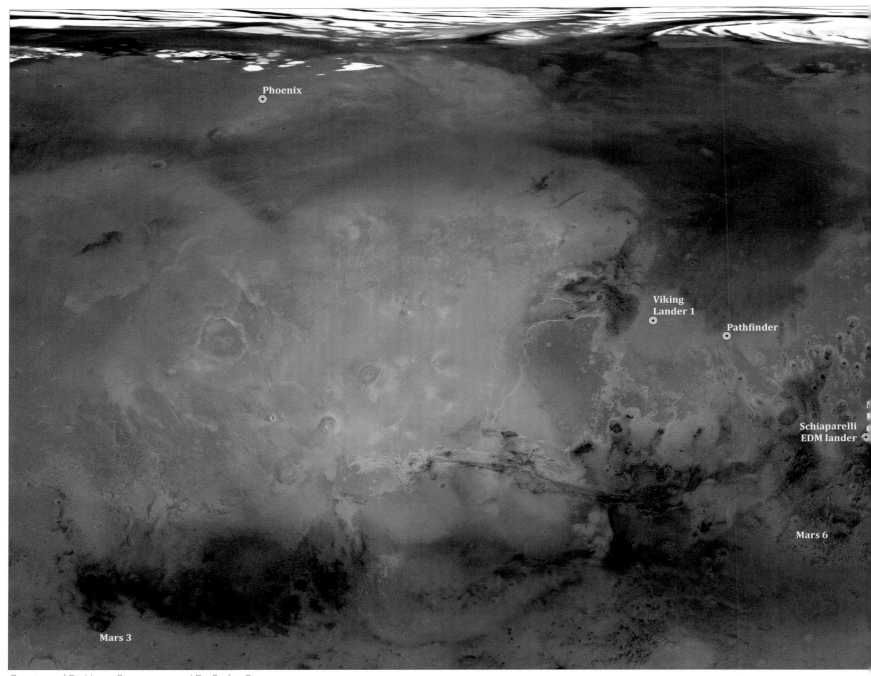

Courtesy of Dr. Vanya Stamenova and Dr. Stefan Stamenov.

HUMAN EXPLORATION OF MARS

Space Research and Technology Institute,
Bulgarian Academy of Sciences
Sofia, Bulgaria
By Vanya Stamenova and Stefan Stamenov

This map identifies the landing sites of various spacecraft sent to explore Mars. The background image is a mosaic of Mars, generated from about 1,000 Viking orbiter images. The map shows the landing sites of the successful missions up to March 2021, as well as the landing sites of unsuccessful missions.

Spacecraft from two successful missions—the

Curiosity rover and InSight lander—are still operational.

On February 18, 2021, the Perseverance rover landed as part of the NASA Mars 2020 mission. The Perseverance landing site is at the Jezero Crater on the western edge of the Isidis Plantina (flat plain). The exact location of the landing is shown with a green circle.

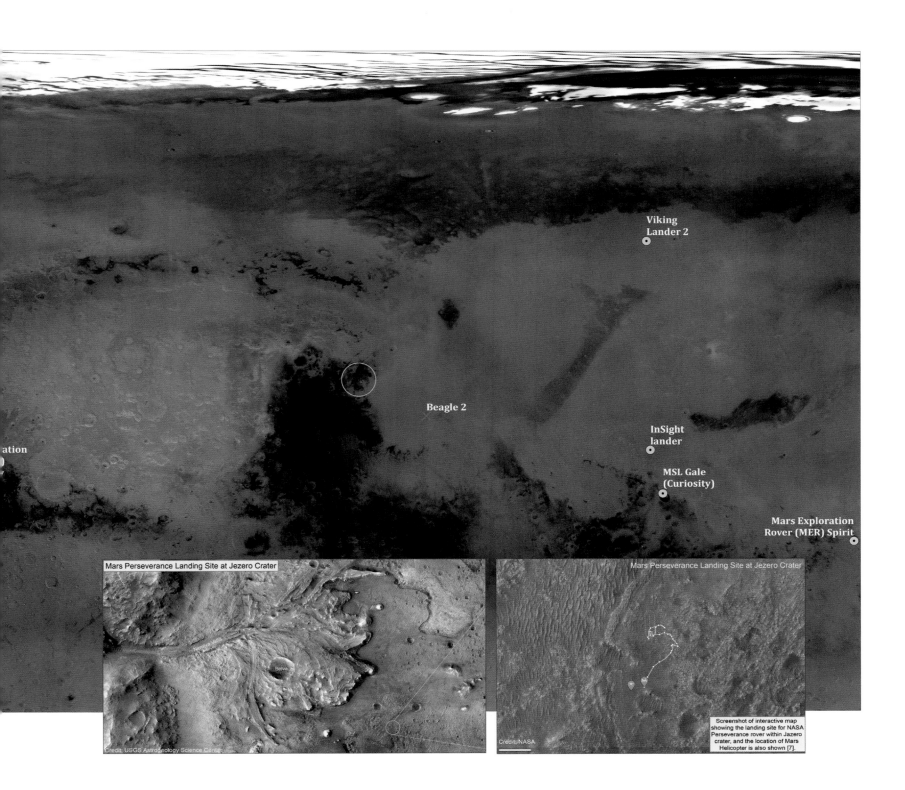

Mars Perseverance Landing Site at Jezero Crater

Mars Perseverance Landing Site at Jezero Crater

Credit: USGS Astrogeology Science Center

Credit: NASA

Screenshot of interactive map showing the landing site for NASA Perseverance rover within Jazero crater, and the location of Mars Helicopter is also shown [7].

CONTACT

Vanya Stamenova
vanya_stamenova@yahoo.com

SOFTWARE

ArcGIS Desktop

DATA SOURCES

USGS Astrogeology Science Center; NASA; NASA History Division;
USGS Astrogeology Science Center; NASA Planetary Data System;
"Mars 2020 Terrain Relative Navigation Flight Product Generation:
Digital Terrain Model and Orthorectified Image Mosaics," 51st
Lunar and Planetary Science Conference (2020); "Context Camera
Investigation on board the Mars Reconnaissance Orbiter," *Journal of
Geophysical Research* (2007)

MARIANA ISLANDS, MICRONESIA

University of Queensland and Island Research & Education Initiative
Hagåtña, Guam
By Maria Kottermair

This map depicts the Mariana Islands and the vast ocean floor surrounding it, including the Mariana Trench, the deepest trench in the world. It provides a better sense of the archipelago's 15 islands in relation to the trench and other ocean features—such as the seamounts that look like stars dotting a night sky. A slight 3D effect was applied to highlight the elevated Mariana Plate that is pushed up by the subducting Pacific Plate, thus creating the Mariana Islands.

This is the first map of its kind showing the topography and bathymetry of that area along with the local CHamoru names of its islands. The islands blend into the surrounding water, showing the strong connection of the CHamoru people with the ocean. The ancient CHamorus were well-known for their seafaring skills and "flying" sail boats that they navigated across the Pacific Ocean.

CONTACT
Maria Kottermair
mariakottermair@gmail.com

SOFTWARE
ArcGIS Pro, Adobe Photoshop

DATA SOURCES
GEBCO 2020; NASA-SRTM; Utugrafihan CHamoru, Guåhan

Courtesy of Maria Kottermair.

Philippine Sea

Pacific
Ocean

Ulåkas

Må'ok

As Songsong

Akligan

Pågan

Alamågan

Guguan

Saligan

Anatåhan

No'os

Sa'ipan

Tini'an

Aguiguan

Luta

Guåhan

Gåni

Låguas

Mariana Trench

EXTRACTING MINING DEPRESSIONS

University of Florida, Gainesville, Florida, USA
By Maram Alrehaili

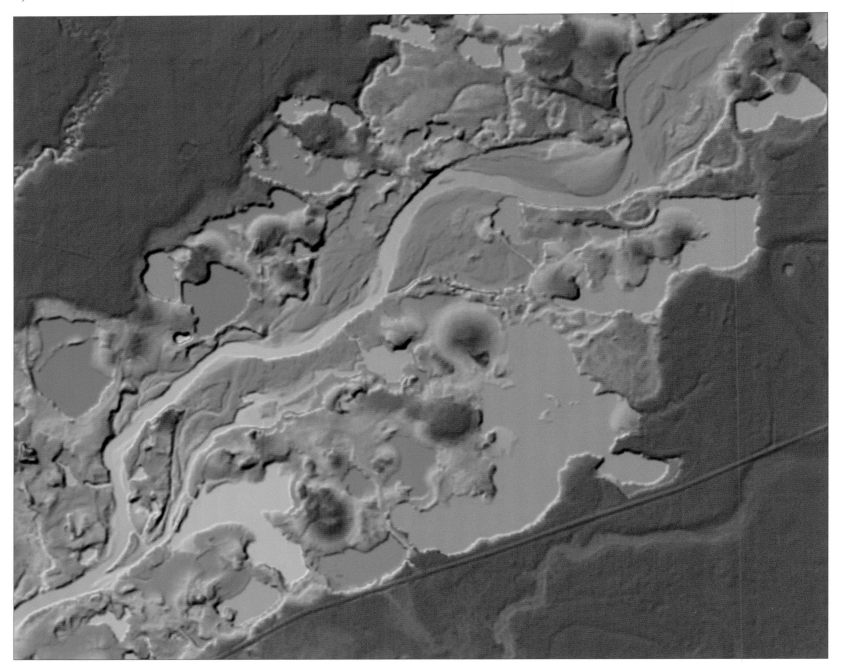

Mining activities in the Middle Amite River are the main cause of channel instability. Therefore, monitoring the interaction between mining activities and the flood plain response is an important task for the state's floodplain management.

This project aims to automate the detection of open pits (which indicates mining) and quantify the nested depression hierarchy of pits. It is expected that there will be more change in pit sizes, avulsions, and configurations in reaches with more floodplain mining. There will be spatial variations in the number and size of pits, coinciding with the intensity of mining.

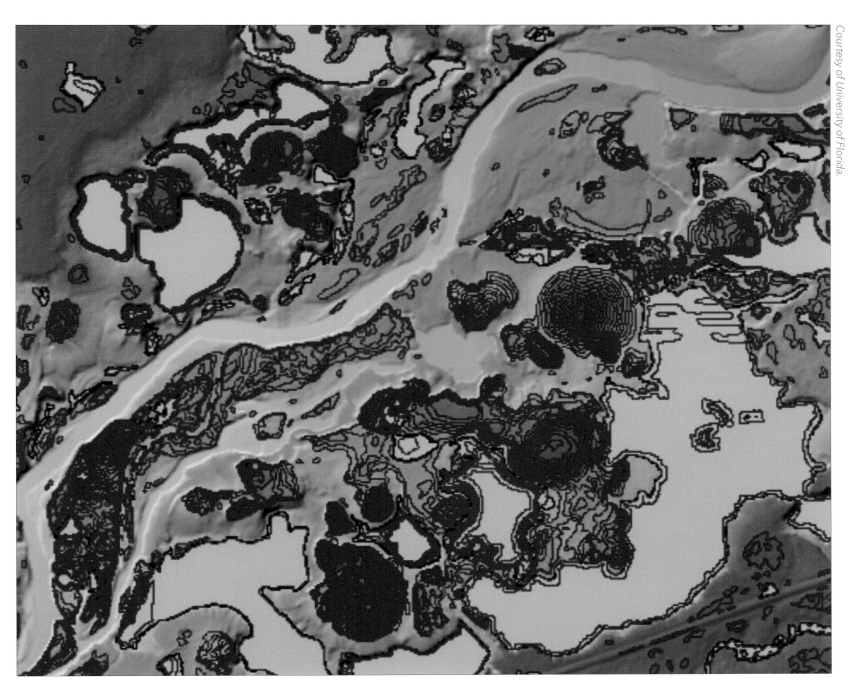

CONTACT
Joann Mossa
mossa@ufl.edu

SOFTWARE
ArcGIS API for Python, ArcGIS
Business Analyst Desktop

DATA SOURCES
NOAA coastal imagery viewer

GROUNDWATER POTENTIAL ZONES IN MATTA, PAKISTAN

Uzair Ahmed
Peshawar, Pakistan
By Uzair Ahmed

This image represents groundwater potential in Matta, Pakistan.

CONTACT

Uzair Ahmed
uzairahmedsparc770@gmail.com

SOFTWARE

ArcGIS Desktop

DATA SOURCES

US Geological Survey Landsat imagery,
Pakistan Irrigation Department

Courtesy of Uzair Ahmed.

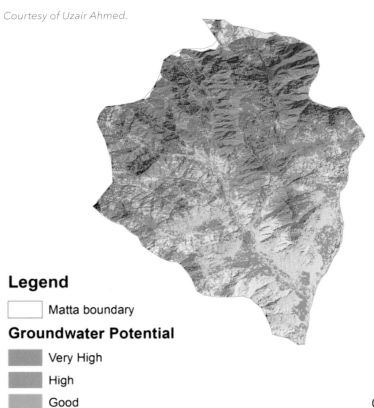

Legend

☐ Matta boundary

Groundwater Potential

Very High

High

Good

Poor

0   2   4   8   12   16
Km

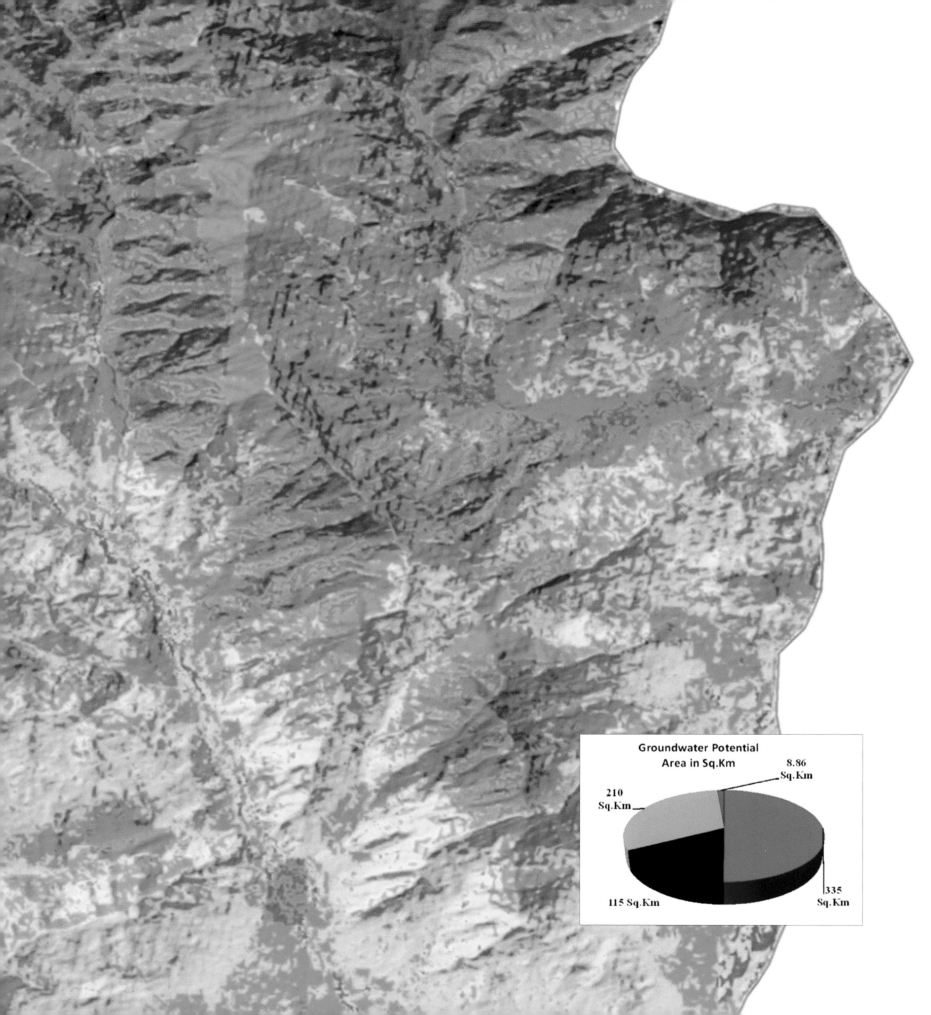

Groundwater Potential
Area in Sq.Km

8.86 Sq.Km

210 Sq.Km

115 Sq.Km

335 Sq.Km

MODELING THE EFFECTS OF A DAM FAILURE

Giorgi Kapanadze
Tbilisi, Georgia
By Giorgi Kapanadze

Among the 17 dams in the country of Georgia, the Enguri Dam stands out. It is an arc-type dam with a height of 271.5 meters.

This model used remote sensing and GIS applications to forecast the impact of a dam failure and identify potential hazards. Estimated flood depths are shown, originating at the dam and diminishing at the Black Sea coast.

CONTACT
Giorgi Kapanadze
kapan.gio777@gmail.com

SOFTWARE
ArcGIS Desktop, QGIS, Envi, Hydrologic Engineering Center's River Analysis System

DATA SOURCES
Remote sensing, Soviet Topographical Maps, National Agency of Public Registry

Courtesy of Giorgi Kapanadze.

Blac

Sea

Legend

● settlement

lake, reservoir

border of municipality

water depth (meters)

| 0 - 5 | 6 - 10 | 11 - 20 | 21 - 40 | 41 - 80 | 81 - 140 |

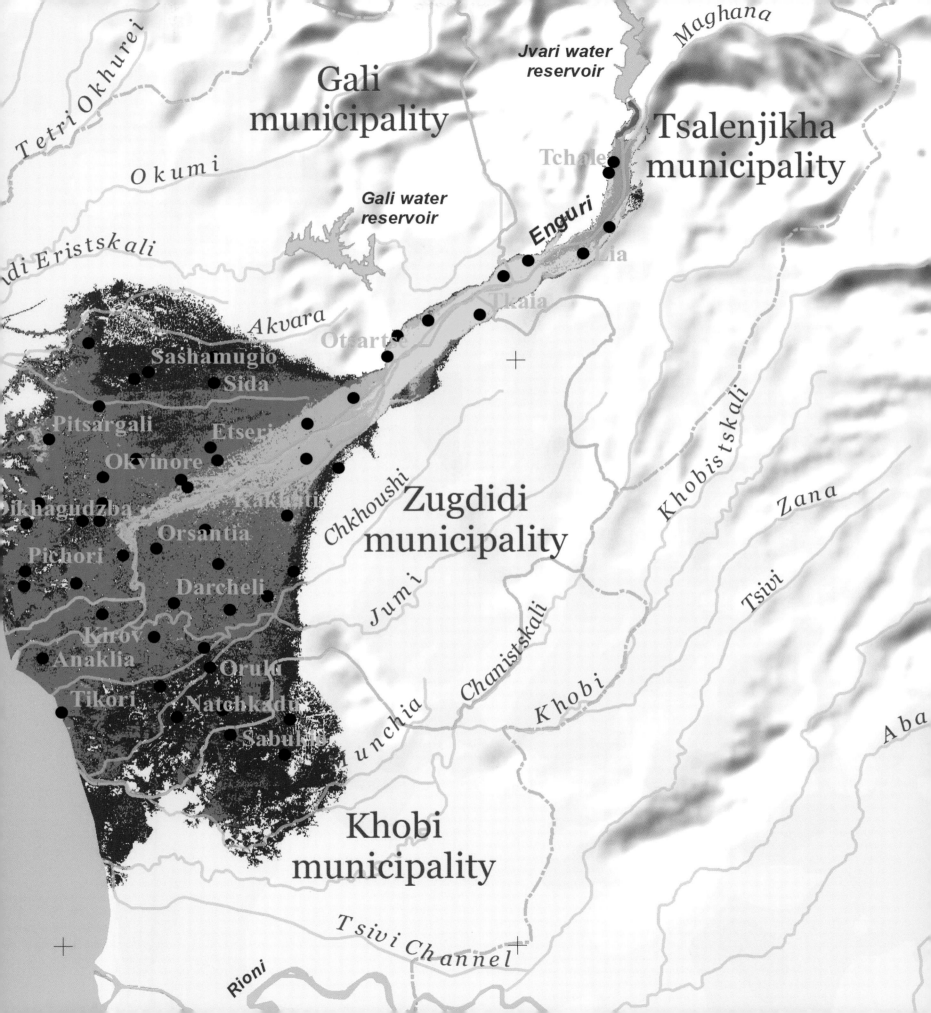

LOST STORAGE CAPACITY IN KANSAS RESERVOIRS

Kansas Water Office
Topeka, Kansas, USA
By Katie Goff, Nathan Westrup, and Richard Rockel

Water supply reservoirs are an important resource for the citizens of Kansas, supplying more than two-thirds of the state's population with drinking water. These reservoirs store water to maintain adequate streamflow through periods of drought and capture flood events, reducing downstream damage. They serve as critical infrastructure, ensuring drought-resilient municipal and industrial water supplies, and additionally provide an important recreational resource to the state. The most significant threat to the useful life of these reservoirs is sediment that gets trapped behind the dam. As sediment accumulates, it compromises storage capacity and supply during future climatic events.

This graphic gives a glimpse into the effects of sedimentation on current reservoir capacity. The vertical height of each bar shows the total storage capacity of a reservoir, the blue portion represents the capacity available in 2021, and brown indicates the volume that has been lost to sediment.

CONTACT
Katie Goff
katie.goff@kwo.ks.gov

SOFTWARE
ArcGIS Desktop, ArcGIS Pro, ArcGIS Spatial Analyst, ArcGIS Living Atlas of the World, ArcScene

DATA SOURCES
Kansas Water Office

Courtesy of Kansas Water Office.

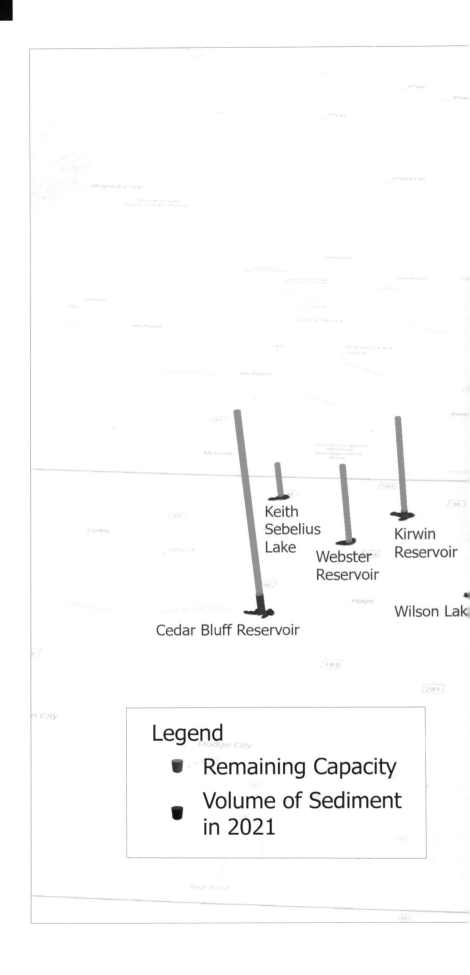

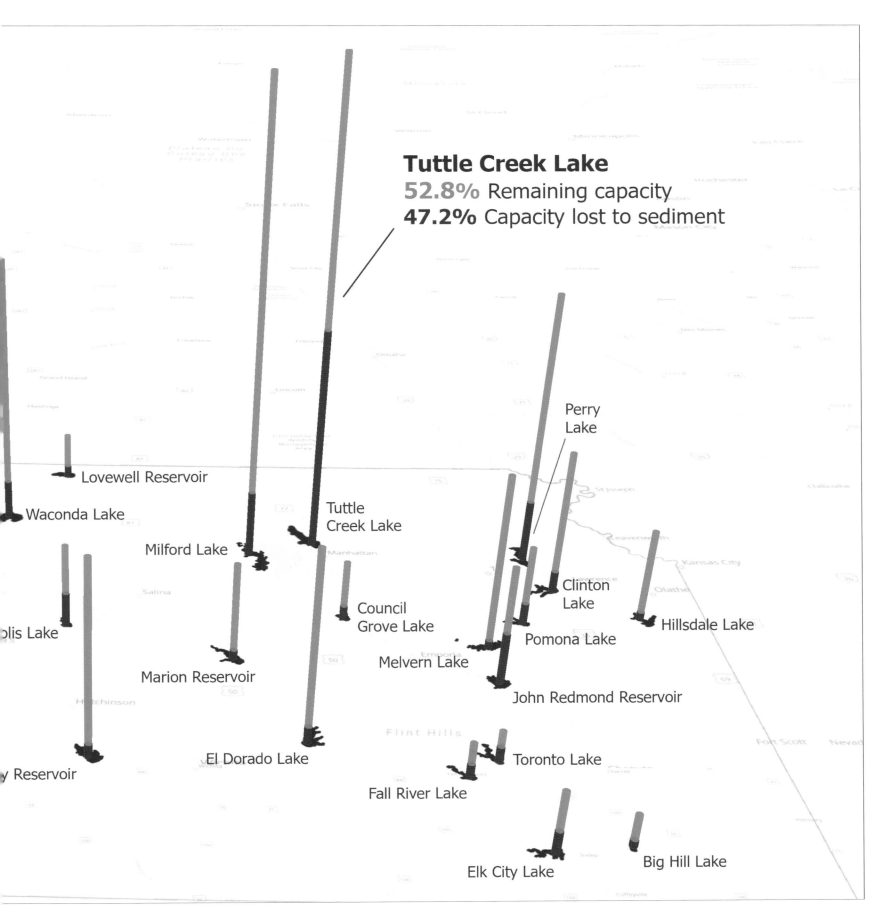

Tuttle Creek Lake
52.8% Remaining capacity
47.2% Capacity lost to sediment

Lovewell Reservoir

Waconda Lake

Tuttle Creek Lake

Milford Lake

Perry Lake

olis Lake

Council Grove Lake

Clinton Lake

Pomona Lake

Hillsdale Lake

Marion Reservoir

Melvern Lake

John Redmond Reservoir

El Dorado Lake

Toronto Lake

Fall River Lake

y Reservoir

Elk City Lake

Big Hill Lake

MAPPING CETACEAN DENSITY IN THE ATLANTIC AND NORTHERN GULF OF MEXICO

Marine Geospatial Ecology Lab, Duke University
Durham, North Carolina, USA
By Ei Fujioka

Estimating the density of marine mammals plays a crucial role in evaluating population health and promoting conservation efforts. It helps resource managers make decisions and aids researchers in conducting studies. The Marine Geospatial Ecology Lab at Duke University, along with partners, built habitat-based density models for marine mammals. They included aerial and shipboard survey data along with environmental variables, to produce a predicted density of marine mammals in the US Atlantic and northern Gulf of Mexico.

This feature-rich mapping tool allows the user to map the predicted density of one species or multiple species and even compare two maps side by side. This example shows the total density of the seven baleen whales in the Atlantic on the right and the species richness on the left. The different visualizations give the user deeper insights.

CONTACT
Ei Fujioka
efujioka@duke.edu

SOFTWARE
ArcGIS API for JavaScript, ArcGIS Desktop, ArcGIS Image Server, ArcGIS Engine

DATA SOURCES
Marine Geospatial Ecology Lab, Duke University; "Habitat-Based Cetacean Density Models for the US Atlantic and Gulf of Mexico," *Scientific Report 6, Article number: 22615* (2016)

Courtesy of the Marine Geospatial Ecology Lab, Duke University.

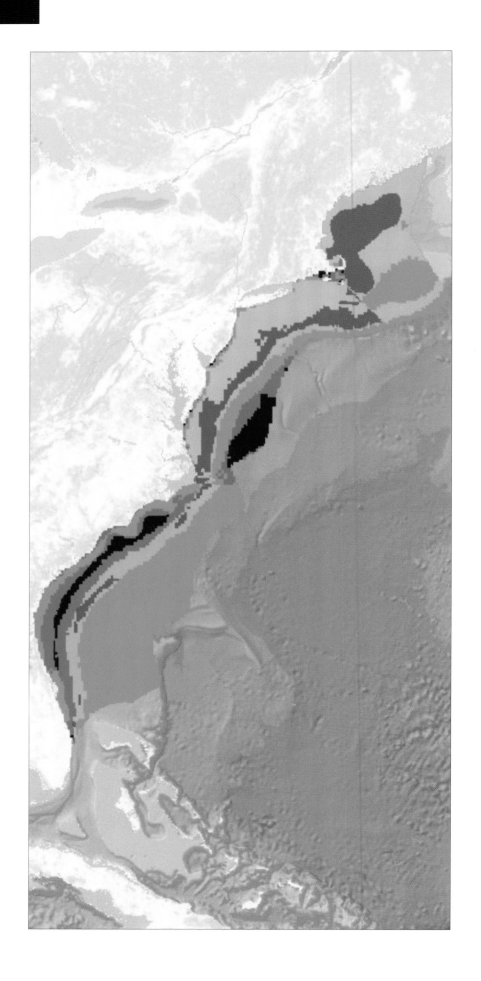

ACTIVE WATER RIGHTS IN THE BAY-DELTA WATERSHED OF CALIFORNIA

California State Water Resources Control Board (SWRCB),
Division of Water Rights
Sacramento, California, USA
By Stanley Mubako

A water right is an authorization to use a reasonable amount of water for a beneficial purpose, and a certain type of water right is required to use any amount of water for domestic use or commercial ventures.

This map shows the types of active water rights for 2018–2019 in the Bay-Delta Watershed, the freshwater lifeline of the state of California. The bulk of water rights are categorized as "Statements of Diversion and Use," followed by "Appropriative," "Registrations," and "Federal Claims."

Identifying the volume and location of water used is crucial for meeting the present and future needs of all Californians, and for protecting the environment from shocks that can arise from unreasonable water diversions or water quality degradation.

CONTACT
Stanley Mubako
smubako@hotmail.com

SOFTWARE
ArcGIS Pro, ArcGIS Desktop extensions,
Microsoft Excel, Inkscape

DATA SOURCES
Basemap and DEM: Esri, DeLorme, NaturalVue, IHO-IOC GEBCO, National Geodetic Survey, USGS, NOAA, Garmin, NPS; water rights data: SWRCB, Division of Water Rights

Courtesy of SWRCB, Division of Water Rights.

LEGEND

Water Right Type

- Appropriative
- Federal Claims
- Registrations
- Statement of Diversion and Use
- Rivers
- Bay-Delta Watershed Boundary

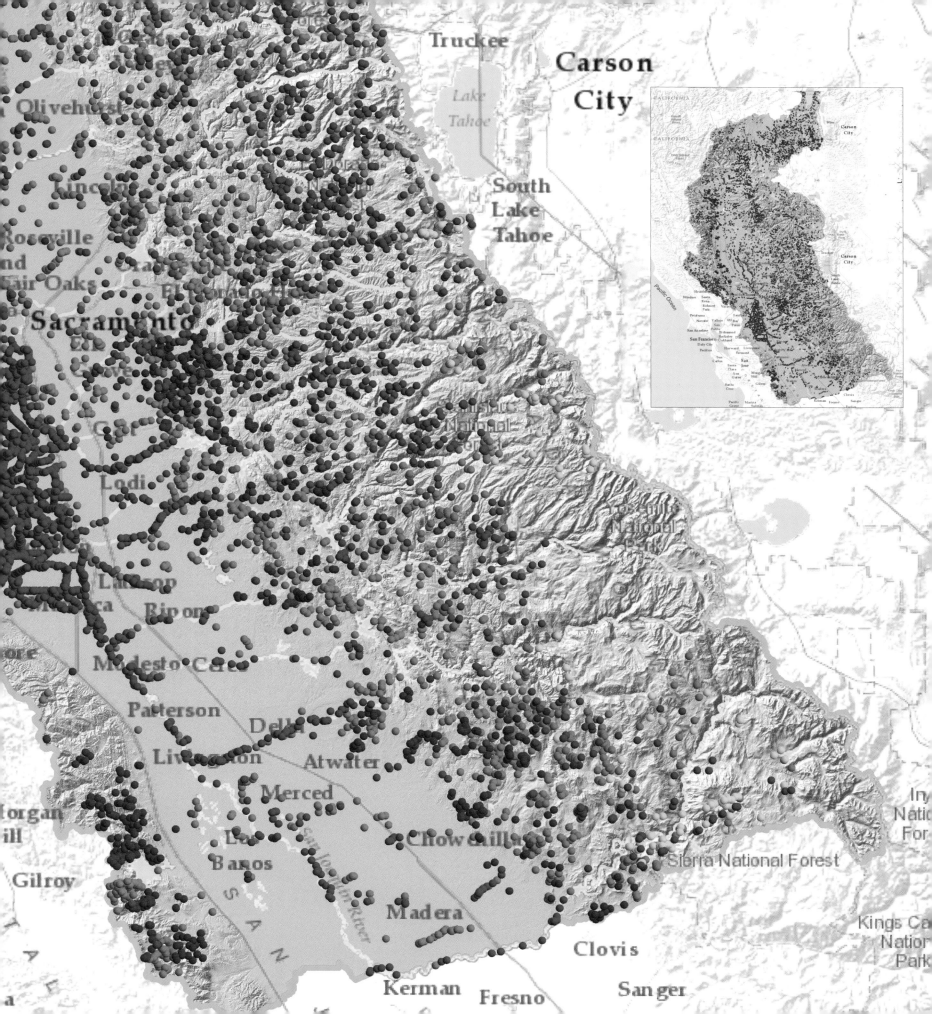

MEASURING THE ANNUAL SHADING IMPACTS OF A BUILDING

University of San Diego, San Diego, California, USA
By Soydan Alihan Polat

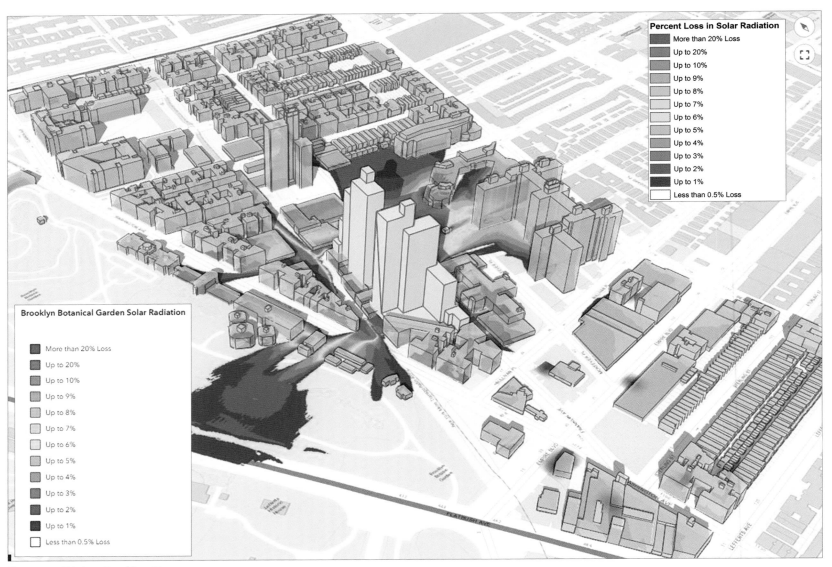

Courtesy of University of San Diego.

Proposed construction of an affordable housing development adjacent to the Brooklyn Botanical Garden required an analysis of its potential to block sunlight from reaching the garden's permanent collections. Net daylight loss is an important metric for analyzing potential impacts on plant health.

Using New York City's 3D Building model and daylight analysis tools, the percentage loss in annual daylight (solar radiation) was calculated and a map depicting that decrease throughout Brooklyn Botanical Garden and the neighboring area was created.

CONTACT
Soydan Alihan Polat
spolat@sandiego.edu

SOFTWARE
ArcGIS Desktop, ArcGIS
Pro, ArcGIS Spatial Analyst,
ArcGIS Web AppBuilder

DATA SOURCES
New York City 3D Building Model (DOITT),
New York City 2' Contours, Environmental
Impact Statement (CEQR # 19DCP095K)

SEISMIC HAZARD SCREENING TO SUPPORT PIPELINE DESIGN IN RURAL AREAS

Tualatin Valley Water District
Beaverton, Oregon, USA
By Scott Fortman

A screening tool for planning a new seismically resilient water supply pipeline was developed that assesses earthquake hazards in rural areas where the topography includes varying slopes and liquefaction potential.

This spatial analysis used lidar data to map the area's topography and determine slope, adding an assessment of the liquefaction risk, derived from existing seismic maps and geotechnical boreholes. Slopes were assigned one of five categories, from low to high. The permanent ground deformation potential was assigned to one of three categories based on the soil's liquefaction displacement index. The combined final score indicated the potential seismic hazard, ranging from low to very high, and informed strategies to withstand a large seismic event.

CONTACT
Scott Fortman
scott.fortman@tvwd.org

SOFTWARE
ArcGIS Pro, ArcGIS Spatial Analyst

DATA SOURCES
Oregon Department of Geology and Mineral Industries (Oregon DOGAMI), Data Resource Center/Metro, Willamette Water Supply Program, McMillian Jacobs Associates

Courtesy of Tualatin Valley Water District.

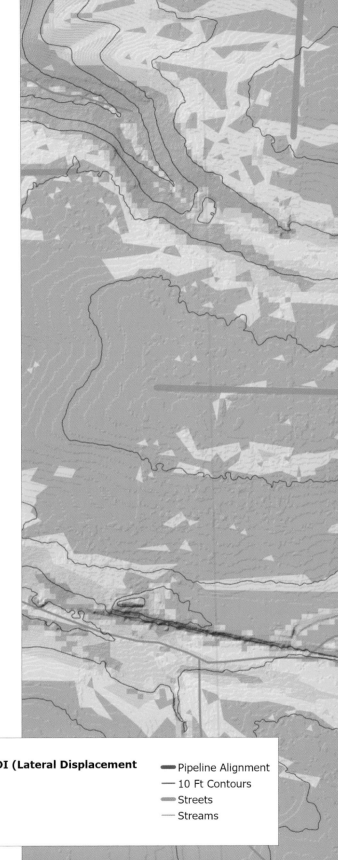

Map Key

Seismic Hazard
Low (0 - 3)
Medium (3 - 6)
High (6 - 12)
Very High (>12)

Boring Site LDI (Lateral Displacement
○ ≤2 ft
◔ ≤6 ft
● ≤15 ft

━ Pipeline Alignment
— 10 Ft Contours
━ Streets
— Streams

0   250   500   1,000 Feet

KURESSAARE CASTLE IN 3D

Estonian Land Board
Kuressaare linn, Saaremaa vald, Saare maakond, Estonia
By Andres Kasekamp

The creation of digital twins has grown rapidly all over the world, and GIS has been the cornerstone of creating and visualizing these models.

This digital twin shows the castle of Kuressaare and its surroundings, extracted from a lidar point cloud. The 3D scene is part of a larger Geo3D strategy whose mission is to model, analyze, and visualize 3D spatial data and make it publicly available. The application integrates buildings, trees, and a digital elevation model.

ONTACT
Andres Kasekamp
Andres.Kasekamp@maaamet.ee

SOFTWARE
ArcGIS Pro, ArcGIS Online, TerraScan

DATA SOURCES
Estonian Land Board

Courtesy of Estonian Land Board.

LAND RECORDS INTERFACE DRIVEN BY USER EXPERIENCE AND FEATURE EXTRACTION EVOLUTION

City of Sioux Falls
Sioux Falls, South Dakota, USA
By Lauri B. Sohl

Land record information is in high demand, both internally and by the public, including residents, developers, consultants, and others. This wide audience requires a robust application in which all types of users can participate in the same arena.

The City of Sioux Falls Parcel Finder application uses the city's published data and takes advantage of numerous widgets. Users can search by address, name, parcel ID, owner name, and alias, using a composite locator. Features can be selected and statistics calculated and exported. The self-service user experience drives the utility and design of this application.

The City of Sioux Falls has an extensive history using imagery to extract planimetric features. The process and products for feature extraction have evolved considerably over the last few years. The City of Sioux Falls has subscribed to annually updated imagery, with periodic traditional ortho and lidar acquisitions. Pretrained deep learning models and the ability to train unique models will provide additional options for creating not only building footprints but numerous other types of surface features.

CONTACT

Lauri B. Sohl
lsohl@siouxfalls.org

SOFTWARE

ArcGIS® Hub℠, ArcGIS Online, ArcGIS Open Data, ArcGIS Pro, ArcGIS Web AppBuilder, ArcGIS Arcade, ArcGIS Image Analyst, ArcGIS Living Atlas of the World, Nearmap

DATA SOURCES

City of Sioux Falls, ArcGIS Living Atlas of the World, Nearmap

Courtesy of City of Sioux Falls.

KENTON COUNTY LAND USE ZONING DEVELOPMENT

Planning and Development Services (PDS) of Kenton County
Covington, Kentucky, USA
By Louis Hill

GIS data is used extensively in the land development decision-making process. Planning and Development Services oversees the land use, zoning, and plat approval processes that shape development in Kenton County, Kentucky.

CONTACT
Louis Hill
lhill@pdskc.org

SOFTWARE
ArcGIS Pro

DATA SOURCES
PDS of Kenton County, LINK-GIS

Courtesy of PDS of Kenton County.

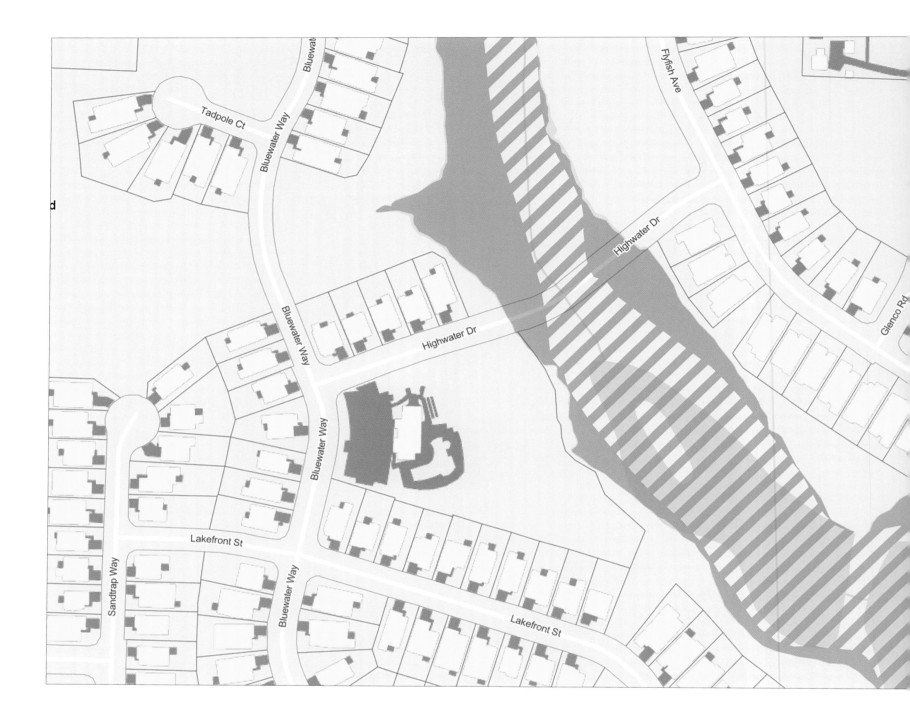

DURHAM MAPS FIT FOR CITY AND COUNTY

City of Durham
Durham, North Carolina,
USA
By Matt Reames

Durham Maps is an application developed by the City of Durham's Technology Solutions Department to support the needs of city, county, and joint city-county departments. By developing an in-house application around the ArcGIS JavaScript API, Technology Solutions can tailor the app's user interface to fit the processes of its business units.

Durham Maps supports both day-to-day staff work and the needs of the public. It includes more than 120 layers from a variety of government sources. Primary uses include address list generation, aerial photos, zoning and planning, parks and trails, permitting and inspections, utilities, the Board of Elections, public schools, economic development, transportation, environmental protection,

AE

AE (Floodway)

Flyfish Ave

fish Ave

Azalea Dr

Impervious Surfaces (Building Footprints)

Impervious Surfaces

- BUILDING
- PAVED
- GRAVEL

FEMA Flood Zones 2018 (for development purposes)

Flood Zones (Development)

- AE (Floodway)
- AE
- AO
- A
- X (1% Future Conditions)
- X (0.2% Annual Chance Flood Hazard)
- X (Area of Minimal Flood Hazard)

Courtesy of City of Durham.

geology, drainage, emergency mitigation, and hazards. This map shows development cases, active building permits the Federal Emergency Management Agency (FEMA) Flood Insurance Rate Map (FIRM) flooding designations..

CONTACT
Matt Reames
matthew.reames@durhamnc.gov

SOFTWARE
ArcGIS Desktop

DATA SOURCES
ArcGIS API for JavaScript, Material Components (MDC) by Google

MAKING A LARGE UNIVERSITY CAMPUS EASIER TO NAVIGATE

Michigan State University
East Lansing, Michigan, USA
By Jade Freeman

Navigating a large university campus can be challenging for visitors, students, and new staff. It can even be difficult for seasoned staff to find locations, assets, and work order sites across campus.

By using features from the exterior road and pathways network, indoor network, parking system, and even parking lot striping, Michigan State University was able to create a network dataset capable of providing specific directions for a variety of travel modes. Staff driving a university-owned vehicle can use the University Vehicle travel mode to generate directions to the nearest parking they are allowed to use and, from there, enter the building and arrive at their destination.

CONTACT
Jade Freeman
Freema23@msu.edu

SOFTWARE
ArcGIS® Indoors™, ArcGIS Network Analyst, ArcGIS Pro

DATA SOURCES
Michigan State University

Courtesy of Michigan State University.

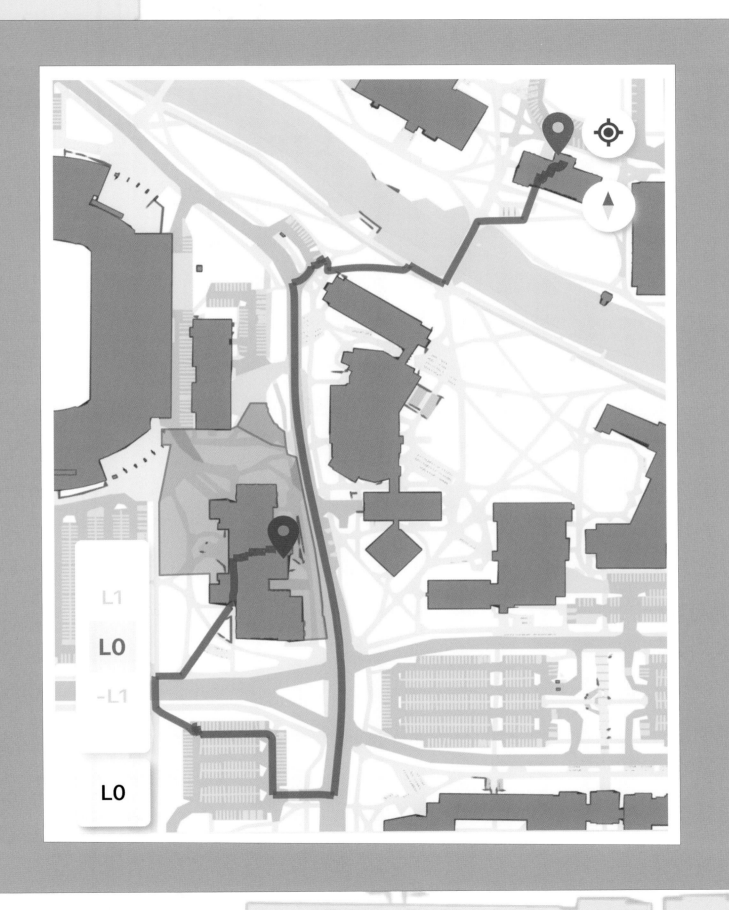

L1

L0

-L1

L0

EXPLORING THE NORTH CHICKAMAUGA CREEK GORGE

University of Tennessee
Chattanooga, Tennessee, USA
By Charlie Mix

North Chickamauga Creek Gorge is located atop the 74-mile long Walden Ridge along the eastern side of the Cumberland Plateau just 20 minutes from Chattanooga, Tennessee, and contains one of the few free flowing streams in the region. Known for its biodiversity, scenic vistas, and world-class outdoor recreation, it is sought out by whitewater kayakers, rock climbers, anglers, and hikers. Since 1989, North Chickamauga Creek Conservancy (NCCC) has worked with state agencies and other nonprofit conservation groups to protect the gorge's unique and threatened ecosystems by conserving more than 17,000 acres of land as state natural area. NCCC continues to advocate for the gorge, its ecosystems, and outdoor recreation.

This map was created to inform people about the gorge and serve as a tool to help the North Chickamauga Creek Conservancy and other organizations continue the stewardship of this southern Appalachian resource. The cartography style pays homage to vintage US Forest Service recreation maps.

CONTACT
Charlie Mix
charles-mix@utc.edu

SOFTWARE
ArcGIS Pro

DATA SOURCES
USGS Protected Areas Dataset, USGS 3D Elevation Program, USGS National Hydrologic Dataset, Tennessee Department of Environment and Conservation, US Census Bureau TIGER Roads, ArcGIS Living Atlas of the World, North Chickamauga Creek Conservancy

Courtesy of University of Tennessee.

Explore North Chickamauga Creek Gorge

Produced by the IGTLab

THE UNIVERSITY OF TENNESSEE CHATTANOOGA

Charlie Mix, GIS Director, 2020

Sources: U.S. Geological Survey, Tennessee Department of Environmental Conservation, U.S. Census Bureau TIGER, Esri, North Chickamauga Creek Conservancy

1 Paradise
Paradise, a beautiful cascading mini-gorge, is a popular spot for hikers and swimmers during the summer. For kayakers, it's one of the areas most infamous class V rapids when its comes out of hibernation during the winter and spring rains.

4 North Chickamauga Creek Falls
A classic Appalachian waterfall that begins with a shallow slide and ends in a twisting torrent. This view is best enjoyed sitting on the river left rocks looking up river as the afternoon sun filters through the mist of the falls.

2 The Rock House
Tennessee's Red Wall Cavern! A beautiful overhanging rock feature next to the recently completed bridge over the Cain Creek segment of the Cumberland trail.

6 The Cumberland Trail
The Cumberland Trail is a long distance hiking trail following some of the most stunning areas along the Cumberland Plateau and many of its gorges. It begins near Chattanooga, TN, and ends at Cumberland Gap National Historical Park in Kentucky. When completed, it will be over 300 miles in length.

3 The Hellican
Named for a nearby highpoint of 1844ft, the Hellican is possibly the most infamous area on Walden Ridge. Throughout the years it has been a destination for swimmers and recreators alike.

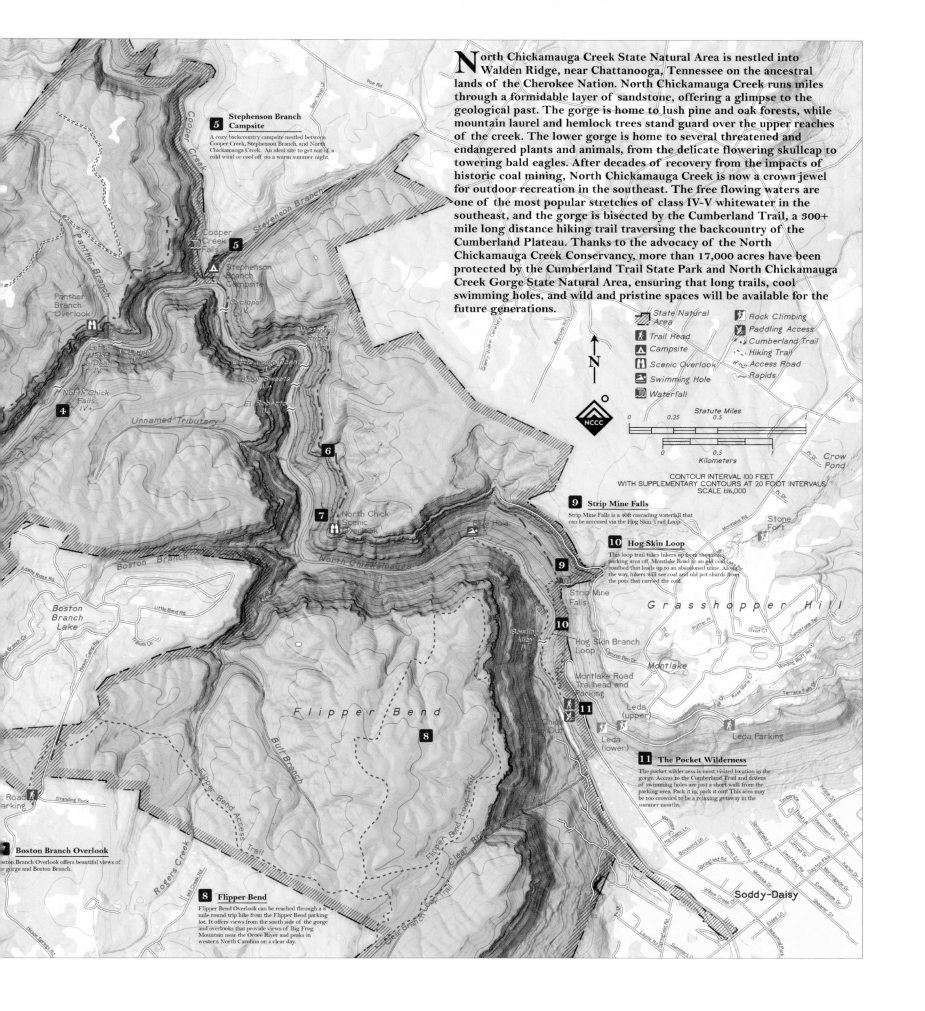

North Chickamauga Creek State Natural Area is nestled into Walden Ridge, near Chattanooga, Tennessee on the ancestral lands of the Cherokee Nation. North Chickamauga Creek runs miles through a formidable layer of sandstone, offering a glimpse to the geological past. The gorge is home to lush pine and oak forests, while mountain laurel and hemlock trees stand guard over the upper reaches of the creek. The lower gorge is home to several threatened and endangered plants and animals, from the delicate flowering skullcap to towering bald eagles. After decades of recovery from the impacts of historic coal mining, North Chickamauga Creek is now a crown jewel for outdoor recreation in the southeast. The free flowing waters are one of the most popular stretches of class IV-V whitewater in the southeast, and the gorge is bisected by the Cumberland Trail, a 300+ mile long distance hiking trail traversing the backcountry of the Cumberland Plateau. Thanks to the advocacy of the North Chickamauga Creek Conservancy, more than 17,000 acres have been protected by the Cumberland Trail State Park and North Chickamauga Creek Gorge State Natural Area, ensuring that long trails, cool swimming holes, and wild and pristine spaces will be available for the future generations.

5 Stephenson Branch Campsite

A cozy backcountry campsite nestled between Cooper Creek, Stephenson Branch, and North Chickamauga Creek. An ideal site to get out of a cold wind or cool off on a warm summer night.

9 Strip Mine Falls

Strip Mine Falls is a 46ft cascading waterfall that can be accessed via the Hog Skin Trail Loop.

10 Hog Skin Loop

This loop trail takes hikers up from the main parking area off, Montlake Road to an old coal roadbed that leads up to an abandoned mine. Along the way, hikers will see coal and old pot shards from the pots that carried the coal.

11 The Pocket Wilderness

The pocket wilderness is most visited location in the gorge. Access to the Cumberland Trail and dozens of swimming holes are just a short walk from the parking area. Pack it in, pack it out. This area may be too crowded to be a relaxing getaway in the summer months.

Boston Branch Overlook

Boston Branch Overlook offers beautiful views of the gorge and Boston Branch.

8 Flipper Bend

Flipper Bend Overlook can be reached through a mile round trip hike from the Flipper Bend parking lot. It offers views from the south side of the gorge and overlooks that provide views of Big Frog Mountain near the Ocoee River and peaks in western North Carolina on a clear day.

Legend

- ▨ State Natural Area
- ☆ Trail Head
- ▲ Campsite
- ⋔ Scenic Overlook
- ⚊ Swimming Hole
- ▥ Waterfall
- ◭ Rock Climbing
- ⚲ Paddling Access
- Cumberland Trail
- Hiking Trail
- Access Road
- Rapids

N

NCCC

Statute Miles

0   0.25   0.5

Kilometers

0   0.5

CONTOUR INTERVAL 100 FEET
WITH SUPPLEMENTARY CONTOURS AT 20 FOOT INTERVALS
SCALE 1:16,000

WHERE DO CHICAGO BASEBALL FANS RESIDE?

SafeGraph
Chicago, Illinois, USA
By Juliana McMillan-Wilhoit

This map shows where people who attend baseball games in Chicago reside. With the use of SafeGraph Patterns, an anonymized and aggregated cell phone mobility dataset, home census block groups are identified for visitors to Wrigley Field in Chicago, Illinois. Visits are counted using SafeGraph's patented machine learning and anonymously assigned to building footprint polygons. In addition to the home census block group, the Patterns dataset provides information such as the volume of visitors to a specific point of interest (POI), time of visit, what other POIs they visit, and more.

CONTACT
Briana Brown
briana@safegraph.com

SOFTWARE
ArcGIS Pro

DATA SOURCES
SafeGraph

Courtesy of SafeGraph.

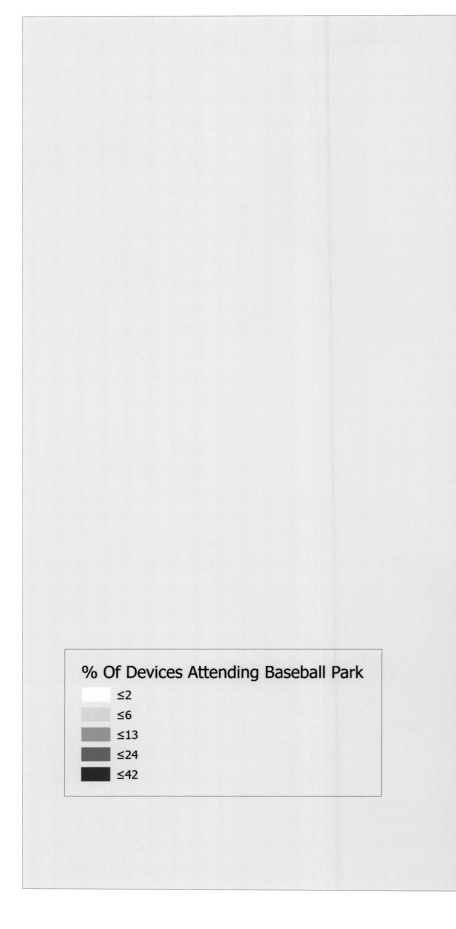

% Of Devices Attending Baseball Park
≤2
≤6
≤13
≤24
≤42

SEQUOIA NATIONAL PARK TRAIL MAP AND GUIDE

USC Spatial Sciences Institute
Los Angeles, California, USA
By Christopher Hayner

This tourist map for Sequoia National Park was designed as part of a graduate seminar. Given the variety of park users and differing interests, the map includes hiking trails, park amenities, and a detail of the major sights around the Giant Forest. The shaded relief map was inspired by the work of Eduard Imhof.

CONTACT
Christopher Hayner
chayner@usc.edu

SOFTWARE
ArcGIS Pro, Adobe Illustrator, Adobe Photoshop

DATA SOURCES
USDA: Natural Resource Conservation Service: Geospatial Data Gateway (National Elevation Dataset, 10 Meter, for Tulare, Fresno, and Inyo Counties; National Hydrology Dataset 1:24,000, for Tulare, Fresno, and Inyo Counties; TIGER streets, for Tulare, Fresno, and Inyo Counties); National Park Service datasets (system boundaries, points of interest, trails and roadways)

Courtesy of USC Spatial Sciences Institute.

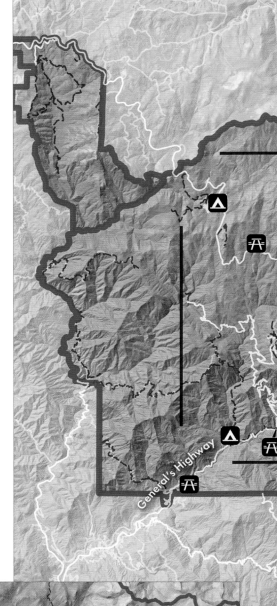

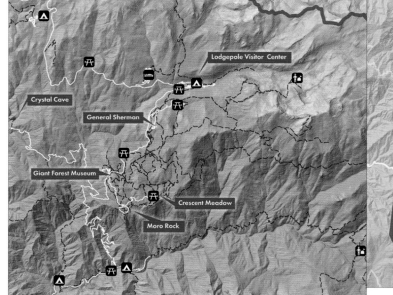

Detail: Giant Forest

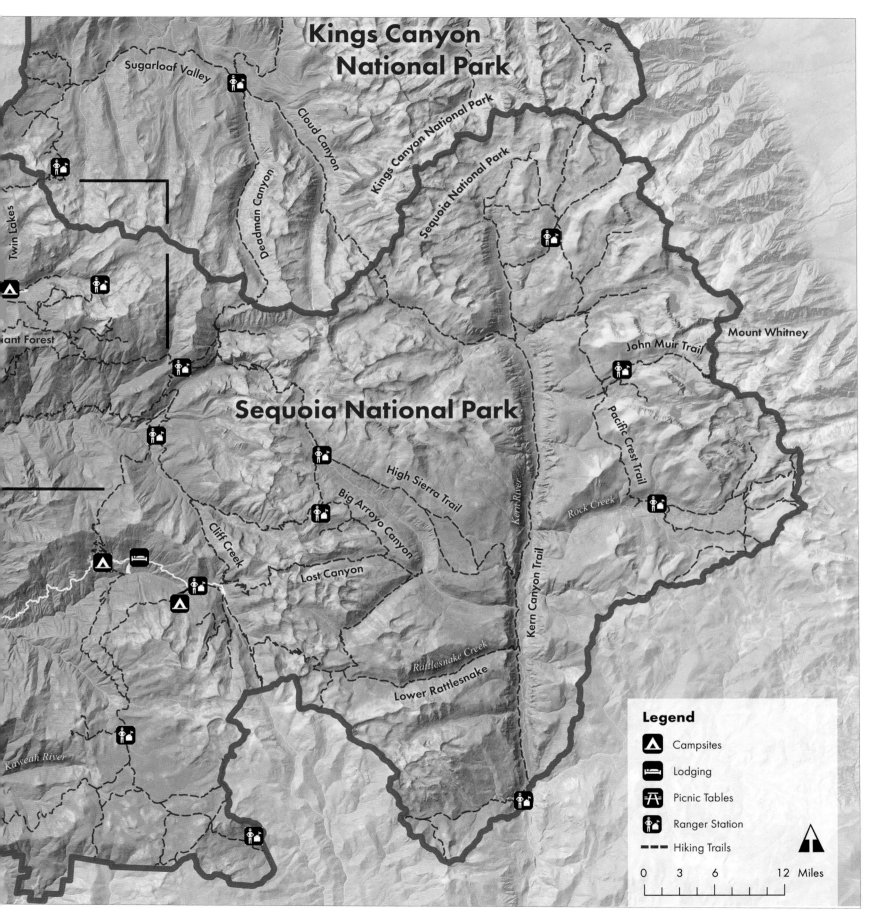

Kings Canyon National Park

Sugarloaf Valley

Cloud Canyon

Kings Canyon National Park

Deadman Canyon

Sequoia National Park

Twin Lakes

Giant Forest

Mount Whitney

John Muir Trail

Sequoia National Park

High Sierra Trail

Big Arroyo Canyon

Pacific Crest Trail

Kern River

Rock Creek

Cliff Creek

Lost Canyon

Kern Canyon Trail

Rattlesnake Creek

Lower Rattlesnake

Kaweah River

Legend

△ Campsites

🛏 Lodging

🪑 Picnic Tables

🧍 Ranger Station

--- Hiking Trails

0   3   6   12   Miles

2021 PARKSCORE INDEX: EQUITY MAPS FOR MINNEAPOLIS

The Trust for Public Land, Santa Fe, New Mexico, USA
By Will Klein, Kirsten Mickow, and Lindsay Withers

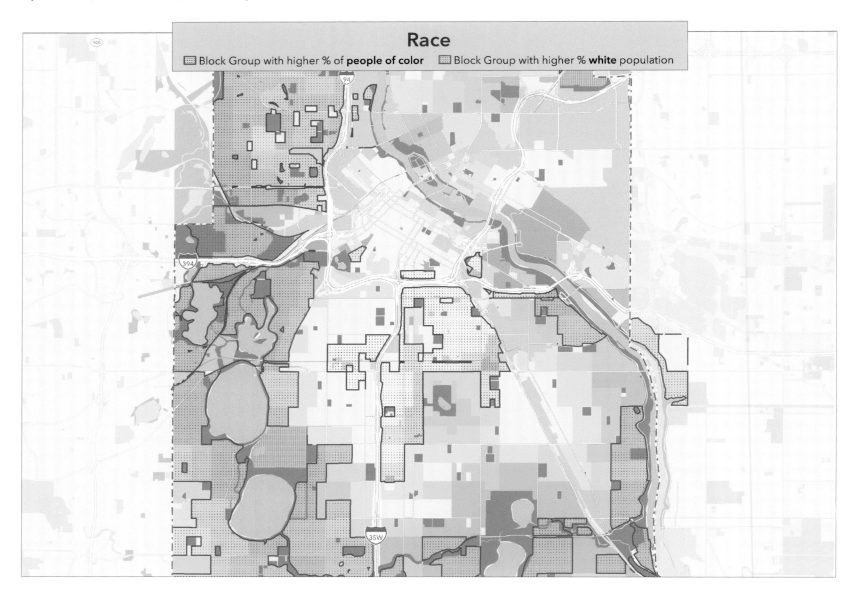

Race

▦ Block Group with higher % of **people of color**   ▦ Block Group with higher % **white** population

For the first time in the ParkScore index's 10-year history, a new measure of equity has been incorporated. It analyzes the amount of neighborhood park space in each of the nation's 100 largest cities by race and income. The data shows significant disparities in accessible park space across racial and economic lines.

This map shows that in Minneapolis, residents in low-income neighborhoods have access to 65% less park space than residents in high-income neighborhoods.

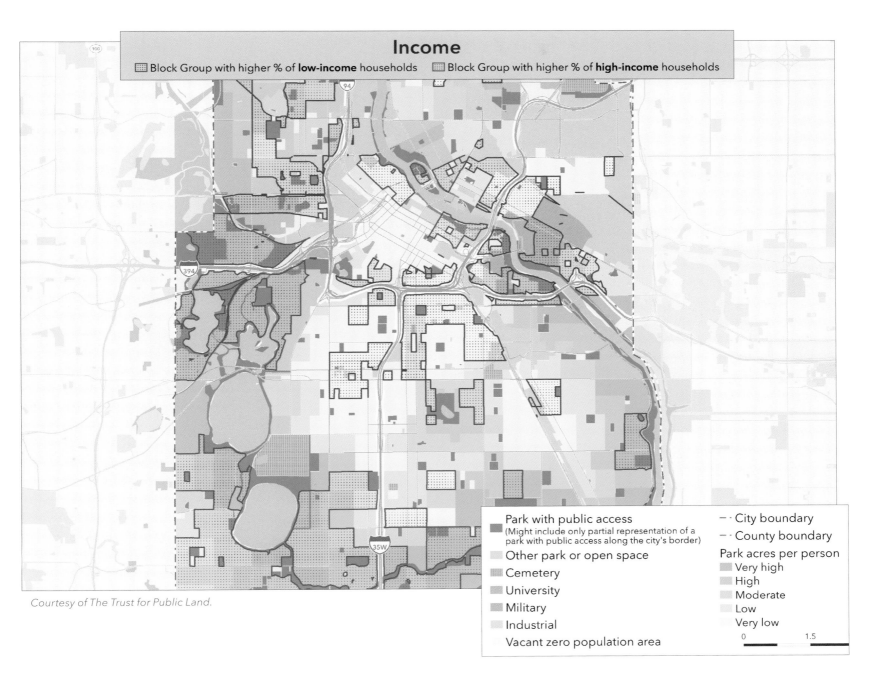

Income

Block Group with higher % of **low-income** households   Block Group with higher % of **high-income** households

Courtesy of The Trust for Public Land.

Park with public access
(Might include only partial representation of a
park with public access along the city's border)

Other park or open space

Cemetery

University

Military

Industrial

Vacant zero population area

— · City boundary

— · County boundary

Park acres per person

Very high

High

Moderate

Low

Very low

0   1.5

CONTACT
Kirsten Mickow
kirsten.mickow@tpl.org

SOFTWARE
ArcGIS Desktop,
ArcGIS Pro

DATA SOURCES
Esri, OpenStreetMap,
City of Minneapolis

WARNER WETLANDS INTERPRETIVE KIOSK

Bureau of Land Management, US Department of Interior
Prineville, Oregon, USA
By USDI Bureau of Land Management

The Warner Wetlands Area of Critical Environmental Concern and Special Recreation Management Area, in South Central Oregon, was designated in 1989 to protect the Warner Valley's unique wetland features, restore critical wildlife habitat, provide high-quality recreation opportunities, and protect other important resource values.

During the last ice age, huge lakes filled Warner Valley. The remnants today form a 40-mile chain of lakes which seasonally flood and recede, depositing sediment on their northeastern shorelines and creating a unique series of bow-shaped dunes. Sweeping vistas of sagebrush spreading across the high desert are punctuated by serene lakes and dominated by Hart Mountain.

Today, this area is visited by thousands of birds during their annual migrations. This habitat is also critical for many other plant, wildlife, and fish species, including the endemic Warner sucker and Warner Lakes redband trout.

This roadside interpretive map attempts to capture the inspiring aesthetic of this remarkable landscape. It was designed to simulate this unique geography in spring when Warner Valley is in bloom and snow caps nearby Hart Mountain.

CONTACT
Gabriel Rousseau
grousseau@blm.gov

SOFTWARE
ArcGIS Spatial Analyst, ArcGIS Desktop, ArcGIS 3D Analyst, Adobe Creative Suite, Avenza Map Publisher

DATA SOURCES
Bureau of Land Management,
US Fish and Wildlife Service, USGS

Courtesy of Bureau of Land Management.

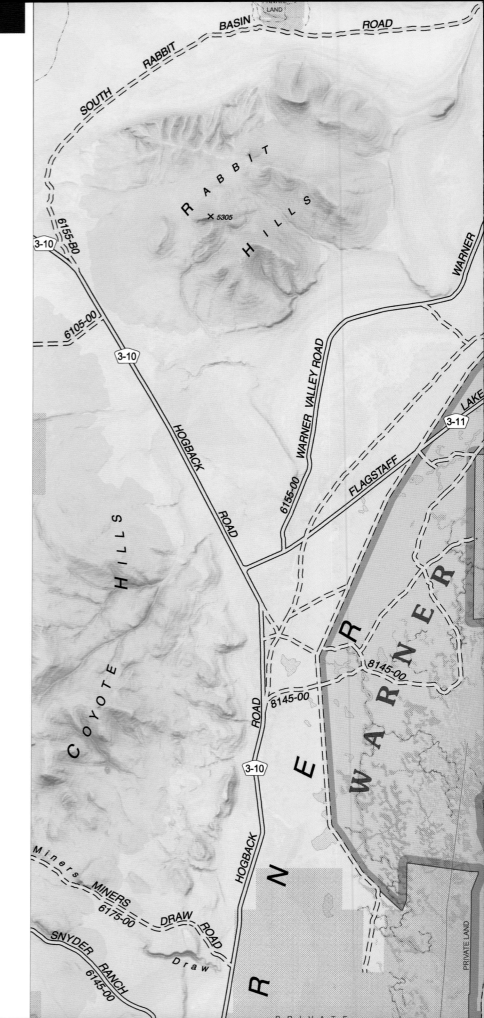

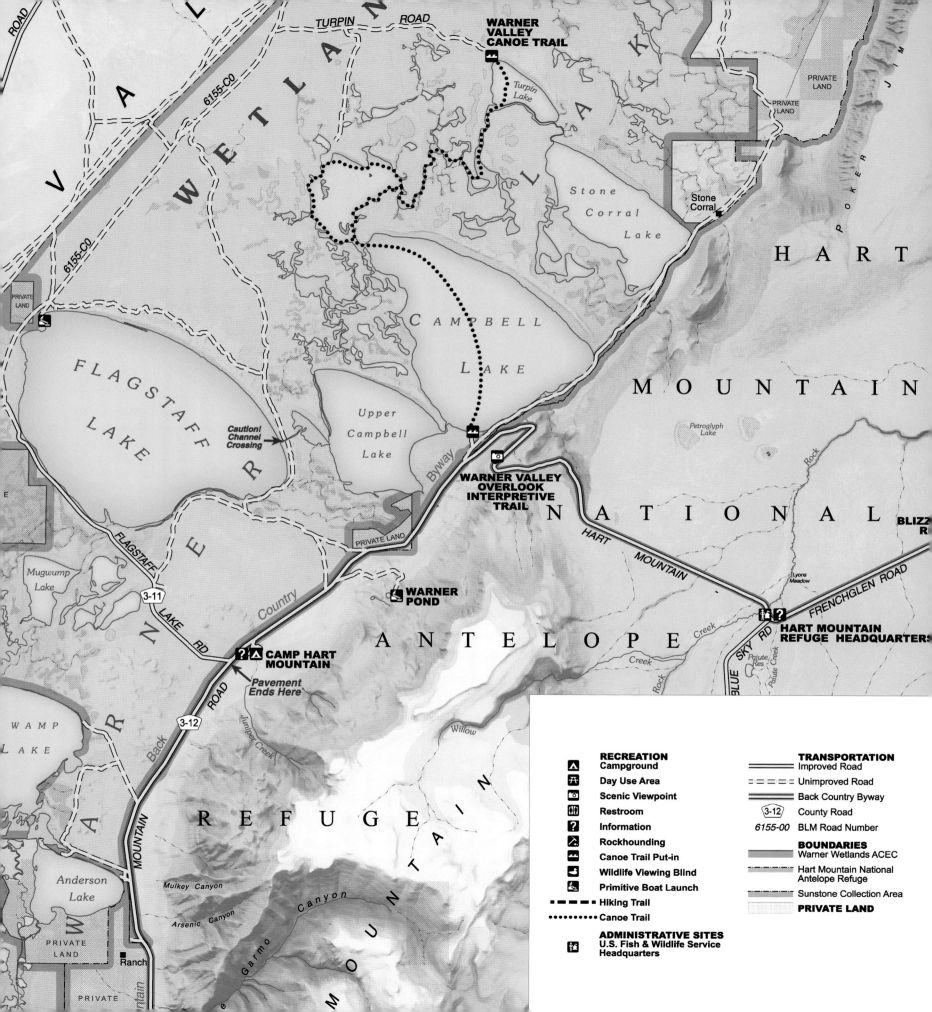

RECREATION

- ▲ Campground
- ⴲ Day Use Area
- ▣ Scenic Viewpoint
- ⑀ Restroom
- ? Information
- ⚒ Rockhounding
- ⮲ Canoe Trail Put-in
- ⛵ Wildlife Viewing Blind
- ⚓ Primitive Boat Launch
- ▪▪▪ Hiking Trail
- •••••• Canoe Trail

ADMINISTRATIVE SITES
U.S. Fish & Wildlife Service
Headquarters

TRANSPORTATION

══════	Improved Road
┅┅┅┅	Unimproved Road
════	Back Country Byway
(3-12)	County Road
6155-00	BLM Road Number

BOUNDARIES

- Warner Wetlands ACEC
- Hart Mountain National Antelope Refuge
- Sunstone Collection Area
- **PRIVATE LAND**

Map labels:
TURPIN ROAD, WARNER VALLEY CANOE TRAIL, Turpin Lake, PRIVATE LAND, PRIVATE LAND, WETLAND, LAKE, HART, Stone Corral Lake, Stone Corral, POKER JIM, 6155-C0, 6155-C0, PRIVATE LAND, V, CAMPBELL LAKE, MOUNTAIN, FLAGSTAFF LAKE, Caution! Channel Crossing, Upper Campbell Lake, Byway, WARNER VALLEY OVERLOOK INTERPRETIVE TRAIL, Petroglyph Lake, NATIONAL, Rock, BLIZZ R, Mugwump Lake, FLAGSTAFF, PRIVATE LAND, HART MOUNTAIN, Lyons Meadow, FRENCHGLEN ROAD, 3-11, LAKE RD, Country, WARNER POND, ANTELOPE, Creek, BLUE SKY RD, Paiute Res, Paiute Creek, HART MOUNTAIN REFUGE HEADQUARTERS, WAMP LAKE, Back, ROAD, Camp Hart Mountain, CAMP HART MOUNTAIN, Pavement Ends Here, MOUNTAIN, Juniper Creek, Rock Creek, Willow, REFUGE, 3-12, Anderson Lake, Mulkey Canyon, Arsenic Canyon, Garmo Canyon, MOUNTAIN, PRIVATE LAND, Ranch, PRIVATE

moveDC: TRANSPORTATION PLANNING IN THE DISTRICT OF COLUMBIA

SymGEO
Washington, DC, USA
By Kevin McMaster

The District Department of Transportation (DDOT) launched an initiative called "moveDC 2021" updating the existing long-range multimodal transportation plan for Washington, DC. To help summarize this report, an ArcGIS map was produced that compiles a wide array of data into a narrative about existing conditions, including current issues and challenges, critical trends, and the state of transportation. This dynamic, interactive, and engaging method of presentation captured the energy and data that has gone into moveDC 2021.

This plan maps mobility networks for bicycles, transit, and freight to achieve mode shift goals and addresses how future mobility trends and innovations will shape the transportation system. It ensures equity is a key consideration in making transportation decisions and engages with the community to reflect current values and meet federal requirements.

CONTACT
Kevin McMaster
kevin.mcmaster@symgeo.com

SOFTWARE
ArcGIS Pro, ArcGIS Online, ArcGIS StoryMaps

DATA SOURCES
Metropolitan Washington Council of Governments

Courtesy of SymGEO.

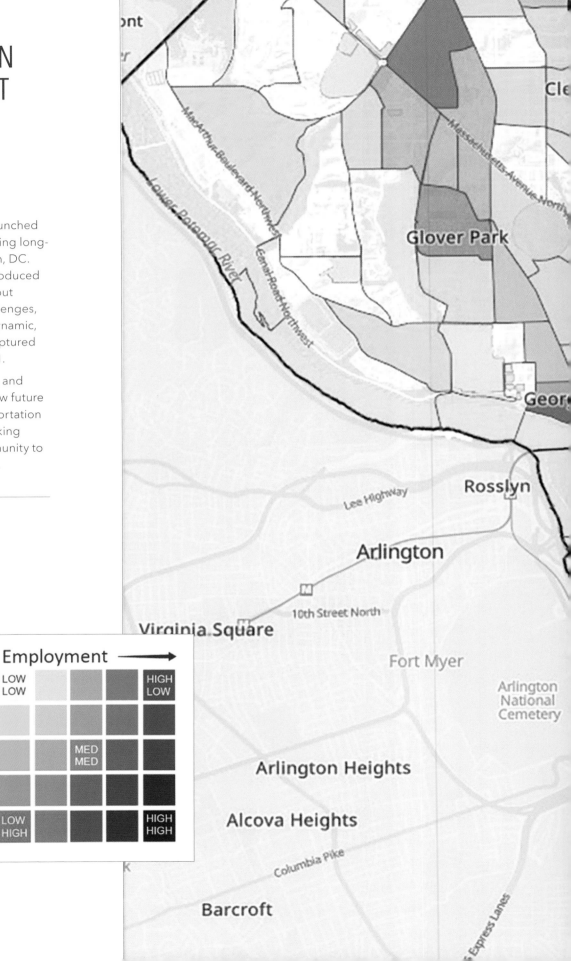

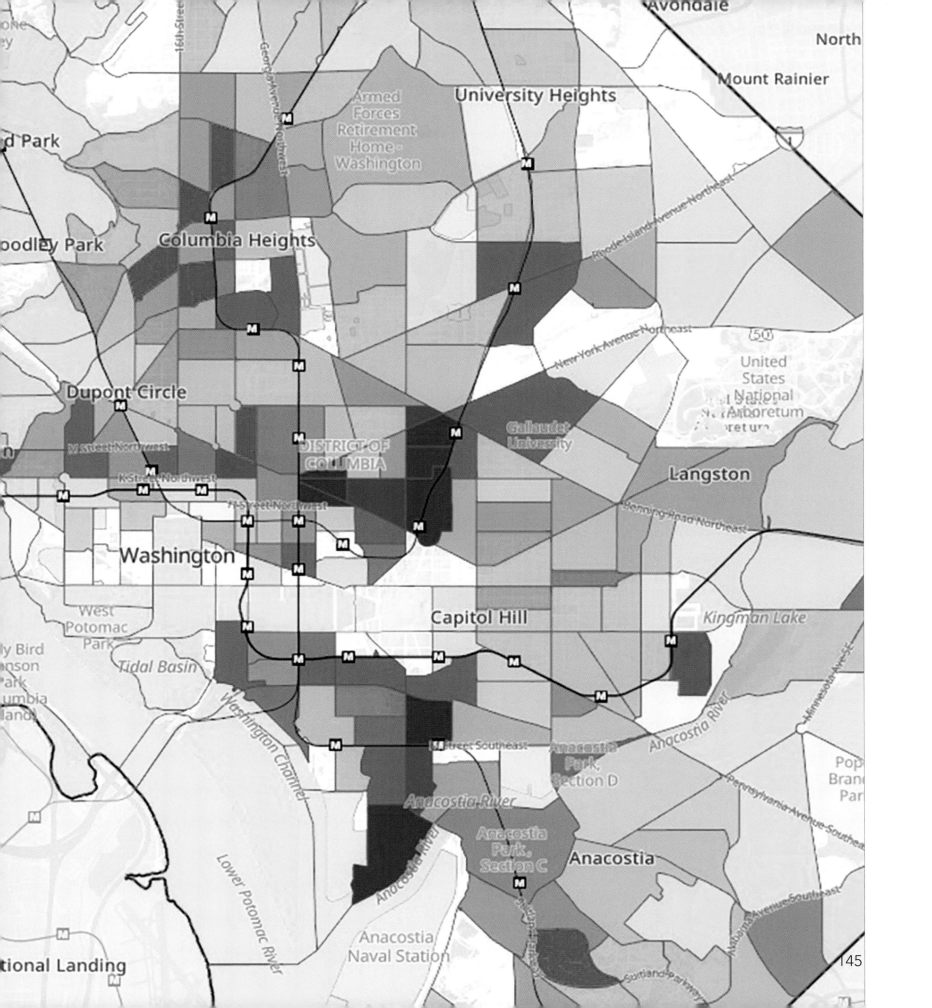

FINDING THE BEST BUS ROUTE USING LOCATION-ALLOCATION ANALYSIS

Warsaw University of Technology
Brzeg Dolny, Poland
By Jakub Kaczorowski

Efficient and accessible public transport is an essential function of a city, enabling its population to fulfil their needs without being forced to use only individual means of transport.

In the city of Brzeg Dolny, an existing bus line was not popular because of its complicated route and long walking distances to bus stops. Data-enriched demand points— residential buildings and points of interest—were determined, and all possible locations of bus stops were automatically created on bus-accessible roads. Using location-allocation analysis to select the bus stops with the shortest distance to demand points, a layer was created with only optimal bus stop locations.

This new distribution of locations will enable 94% of residents to get to a bus stop within five minutes of walking, in comparison with 55% for the previous stop locations. This map shows the proposed bus route serving the optimized bus stop locations.

CONTACT
Jakub Kaczorowski
jkkaczorowski@gmail.com

SOFTWARE
ArcGIS Pro

DATA SOURCES
Local Data Bank of Poland, OpenStreetMap data

Courtesy of Warsaw University of Technology.

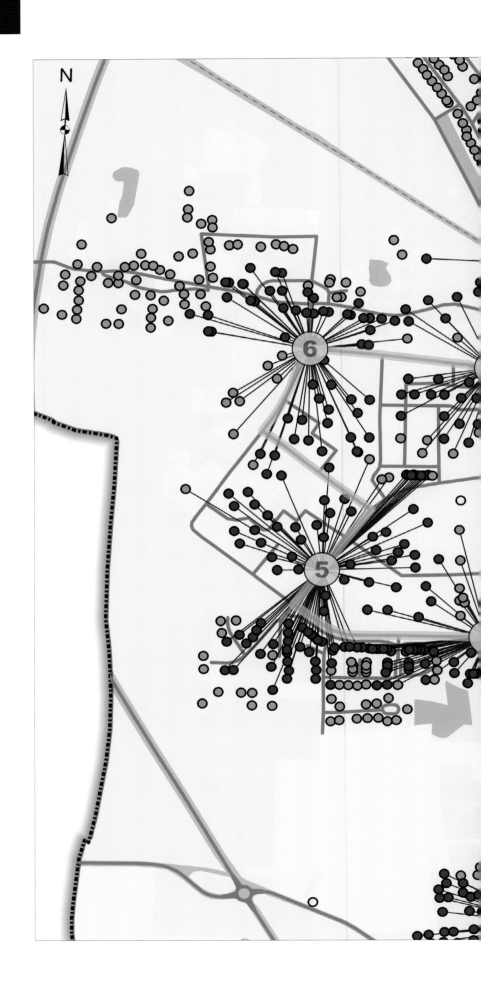

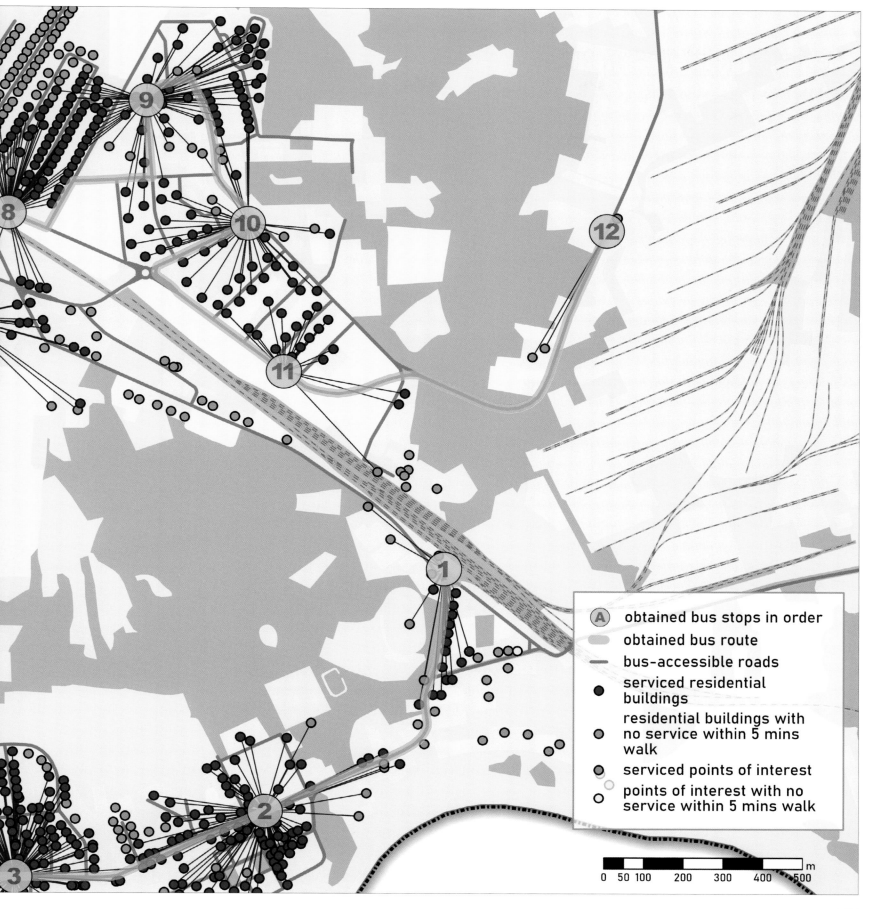

(A)	obtained bus stops in order
	obtained bus route
	bus–accessible roads
●	serviced residential buildings
●	residential buildings with no service within 5 mins walk
●	serviced points of interest
○	points of interest with no service within 5 mins walk

0 50 100   200   300   400   500 m

147

COLLISIONS ON THE OTTAWA NETWORK

Jan Acusta
Ottawa, Ontario, Canada
By Jan Acusta

The City of Ottawa publishes collision data, road center lines, and neighborhood boundaries in its Open Ottawa data portal. This map provides an overview of the number of collisions per road segment in Centretown and its six surrounding neighborhoods.

In this analysis, collision datasets from 2014 to 2018 were merged and shown on the road network. The road center lines were symbolized to resemble two-way traffic, with neighborhood boundaries highlighting the area of interest and adding texture to the map.

CONTACT
Jan Acusta
jan.acusta@outlook.com

SOFTWARE
ArcGIS Pro

DATA SOURCES
Open Ottawa

Courtesy of Jan Acusta.

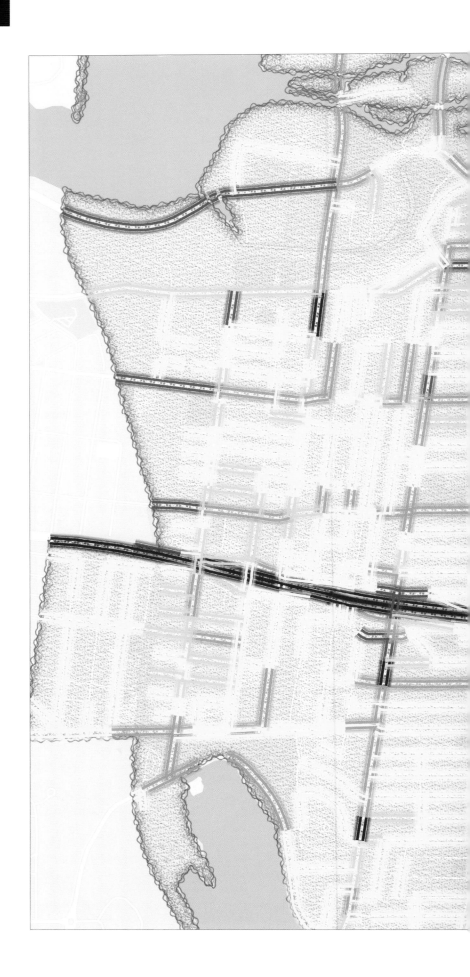

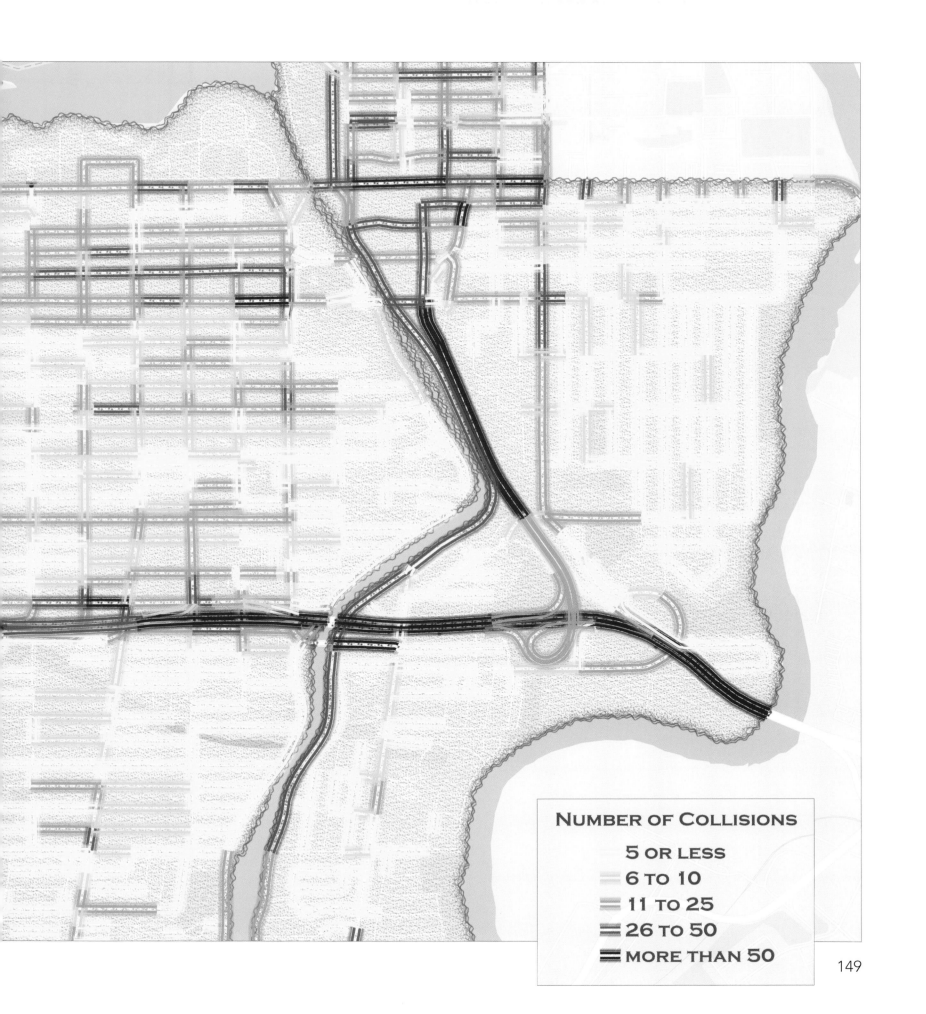

NUMBER OF COLLISIONS

5 OR LESS
6 TO 10
11 TO 25
26 TO 50
MORE THAN 50

149

TRANSPORTATION OPTIONS FOR LOS ANGELES COUNTY METRO

DCR Design LLC
Redlands, California, USA
By Georgia Crowley and Nils Figueroa

This map, created for the Los Angeles County Metropolitan Transportation Authority (Metro), shows rail and core bus routes as well as important landmarks in Los Angeles County. It allows users of public transportation to plan trips that use any of the region's multiple transit agencies. It was created by integrating bus routes with OpenStreetMap and other GIS data sources. The data was simplified and revised before moving it to Adobe Illustrator for final cartographic editing, generalization, and the addition of graphic illustrations.

CONTACT

Roland Hansson
roland@dcrdesign.net

David Figueroa
david@dcrdesign.net

SOFTWARE

ArcGIS Desktop, Adobe Illustrator

DATA SOURCES

Los Angeles Metro Google Transit Feed Specification and other GIS layers, OpenStreetMap

Courtesy of Los Angeles Metro and DCR Design LLC.

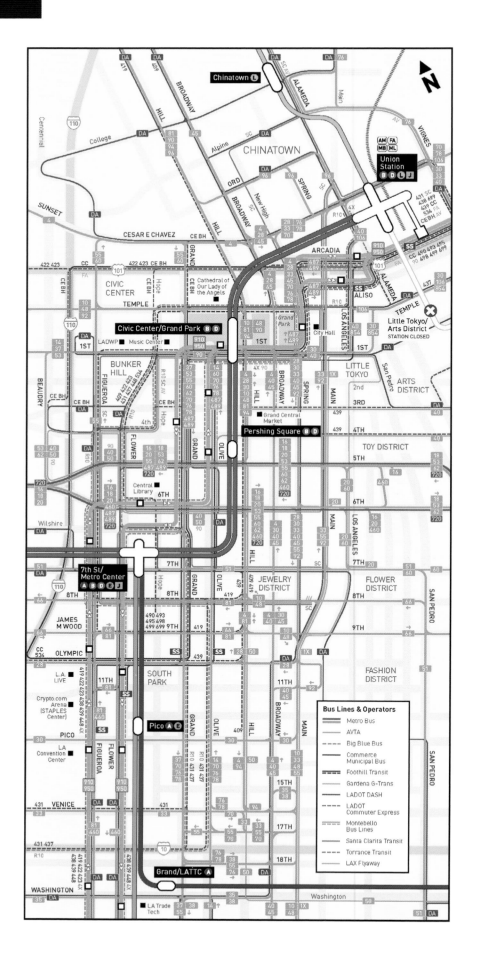

COMMUTE TIMES IN MARYLAND COUNTIES

Maryland Department of Transportation, Secretary of Transportation's Office
Hanover, Maryland, USA
By Andrew Bernish

Maryland borders several states and the District of Columbia, so residents often travel outside of both their state and county for work. The Maryland Department of Transportation (MDOT) showed this map in a presentation discussing economic development strategies for Maryland counties and used it to encourage and challenge counties to prioritize their own residents as they work to attract new job opportunities.

Two southern Maryland counties, neither of which border the District of Columbia or a different state, have the longest commute times with average commutes of greater than 40 minutes. Charles County, with a nearly 45-minute commute, has more than 70% of its resident labor force commuting outside the county.

CONTACT
Andrew Bernish
abernish@mdot.maryland.gov

SOFTWARE
ArcGIS Pro, ArcGIS Maps for Adobe Creative Cloud, Adobe Illustrator, Microsoft Excel, ArcGIS Vector Tile Style Editor

DATA SOURCES
Maryland Department of Transportation, US Census Bureau American Community Survey

Courtesy of Maryland Department of Transportation, Secretary of Transportation's Office.

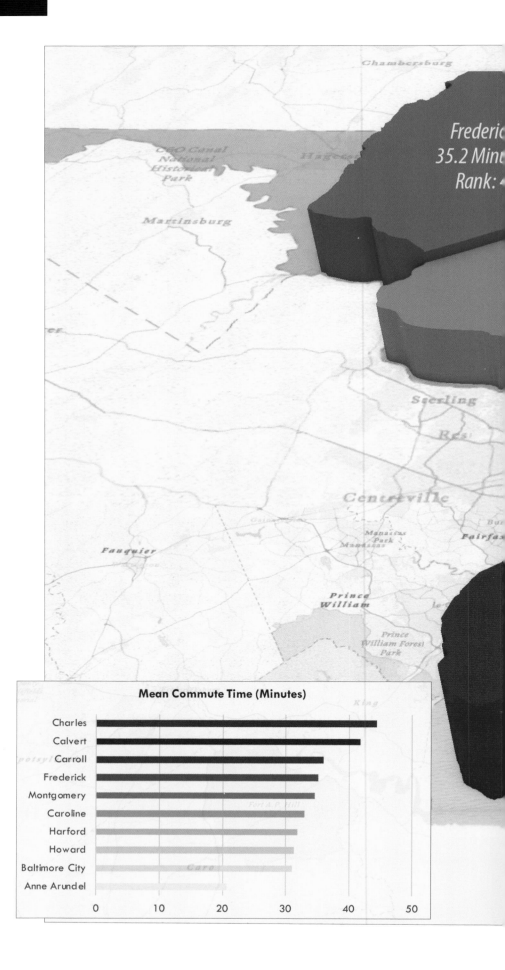

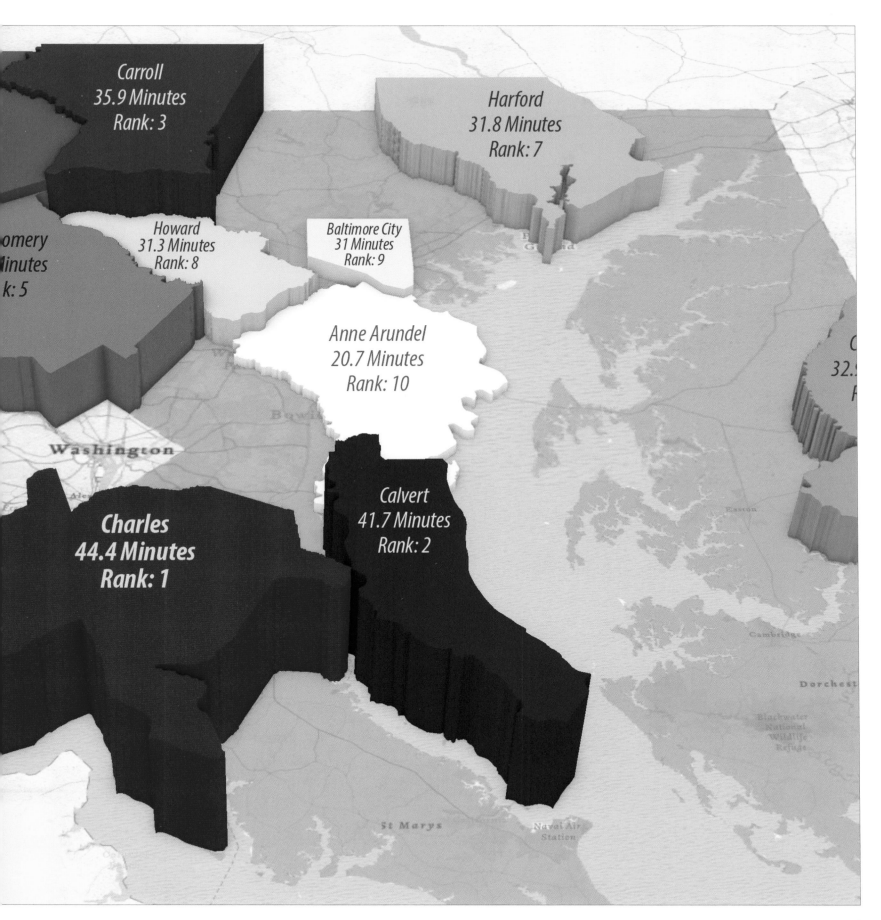

Carroll
35.9 Minutes
Rank: 3

Harford
31.8 Minutes
Rank: 7

omery
Minutes
k: 5

Howard
31.3 Minutes
Rank: 8

Baltimore City
31 Minutes
Rank: 9

Anne Arundel
20.7 Minutes
Rank: 10

C
32.
R

Washington

Charles
44.4 Minutes
Rank: 1

Calvert
41.7 Minutes
Rank: 2

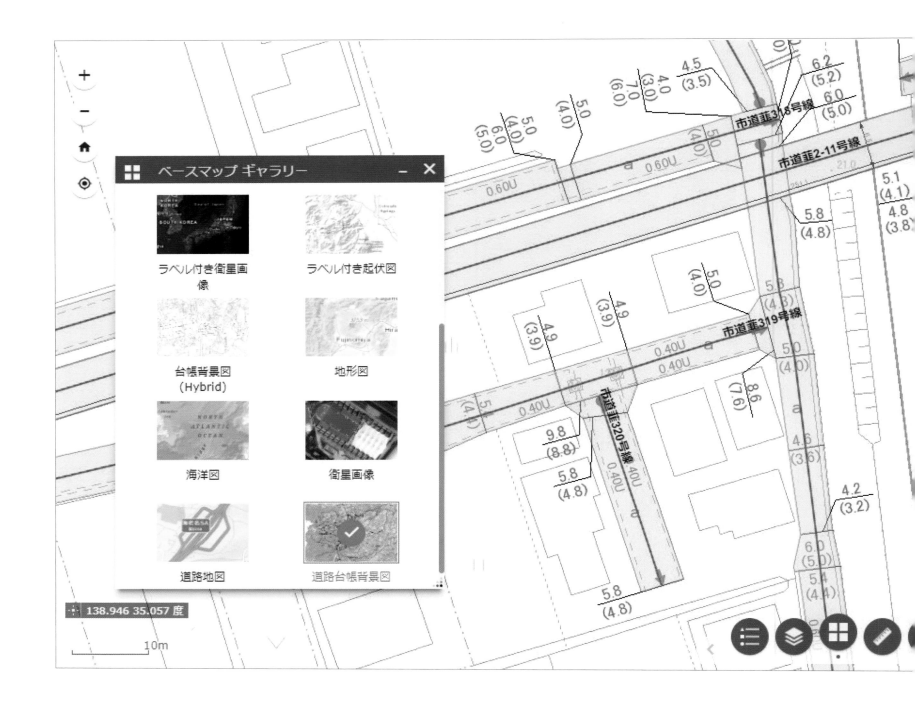

ROAD LEDGER SYSTEM IN IZUNOKUNI CITY

Izunokuni City
Shizuoka, Japan
By Izunokuni City

The City of Izunokuni, Japan, has made its road system available online to residents, who can view it and use it to report issues they encounter.

On high-precision topographic maps created using AutoCAD and ArcMap™, polygons were overlaid indicating road areas managed by the local government, and annotation layers were added describing road information including width, curvature, and more.

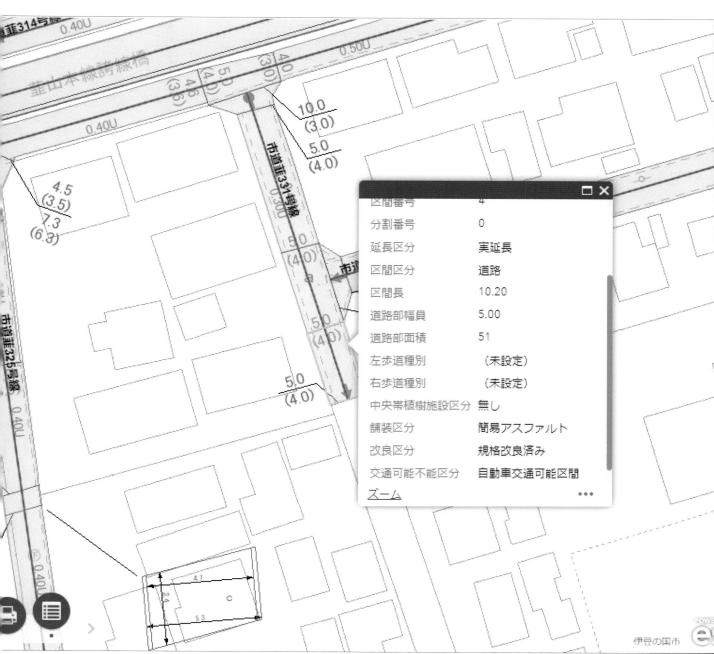

Courtesy of Izunokuni City.

CONTACT
Tohnichi Co. Ltd.
soumu@tohnichi-net.co.jp

SOFTWARE
ArcGIS Online,
ArcGIS Desktop,
ArcGIS Web AppBuilder

DATA SOURCES
Izunokuni City

ASSESSING TRAIL ACCESS AND EQUITY IN THE UNITED STATES

Rails to Trails Conservancy
Washington, DC, USA
By Derek Strout

Trails and greenways have the potential to deliver powerful benefits to communities—providing people of every age, race, ability, and socioeconomic background safe and inexpensive spaces for outdoor physical activity, commuting, and recreation. Trails can also serve as economic catalysts and provide critical "social infrastructure"—public spaces where people can meet, interact, and build relationships.

This image displays results of a preliminary analysis of where trails exist and, perhaps more importantly, where they don't. The map attempts to identify census tracts that lack trail infrastructure and whose community demographic and health characteristics indicate a potentially higher level of impact if they were to receive new infrastructure investments.

By combining a nationwide trail dataset with robust statistics from the CDC's Social Vulnerability Index and the EPA's Enhanced Qualified Opportunity Zones and providing easy-to-use filtering and querying tools, the map offers an interactive and customizable interface that can be valuable for diverse audiences. Community members, trail advocacy organizations, and policy makers at all levels of government can use these maps to help identify areas of opportunity in their local areas to build, expand, and connect equitable and inclusive trails and trail networks.

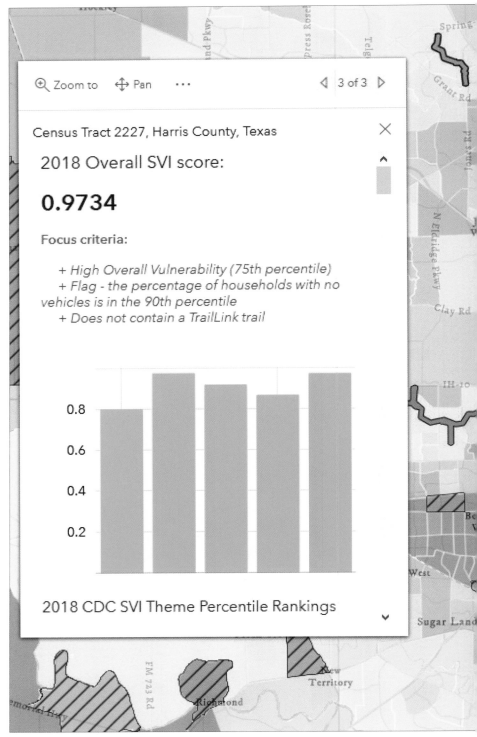

Courtesy of Rails to Trails Conservancy.

EPA-Enhanced Qualified Opportunity Zones (tracts)

CDC SVI greater than .75 (0-1)
EPA Walkability less than 10 (0-20)
Sorted by 2018 population
Click to zoom

Osceola County, Florida
Tract 12097042900
ACS 2018 Pop: 21,843
CDC SVI: 0.831
EPA Walkability: 8.500

Osceola County, Florida
Tract 12097041100
ACS 2018 Pop: 20,419
CDC SVI: 0.851
EPA Walkability: 9.083

Hidalgo County, Texas
Tract 48215024201
ACS 2018 Pop: 15,877
CDC SVI: 0.966
EPA Walkability: 2.944

Hidalgo County, Texas
Tract 48215021302
ACS 2018 Pop: 15,786
CDC SVI: 0.904
EPA Walkability: 6.889

Washington County, Arkansas
Tract 05143010302
ACS 2018 Pop: 15,534
CDC SVI: 0.859
EPA Walkability: 7.667

Montgomery County, Texas
Tract 48339692602

◄ EPA-EQOZ ►

8,062,391 people

live in EPA-EQOZs with high SVI and low walkabilit

CONTACT
Derek Strout
derek@railstotrails.org

SOFTWARE
ArcGIS Dashboards,
ArcGIS Living Atlas of the World,
ArcGIS Online, ArcGIS Pro

DATA SOURCES
Esri, US Centers for Disease Control and Prevention (CDC),
US Environmental Protection Agency (EPA), OpenStreetMap
contributors

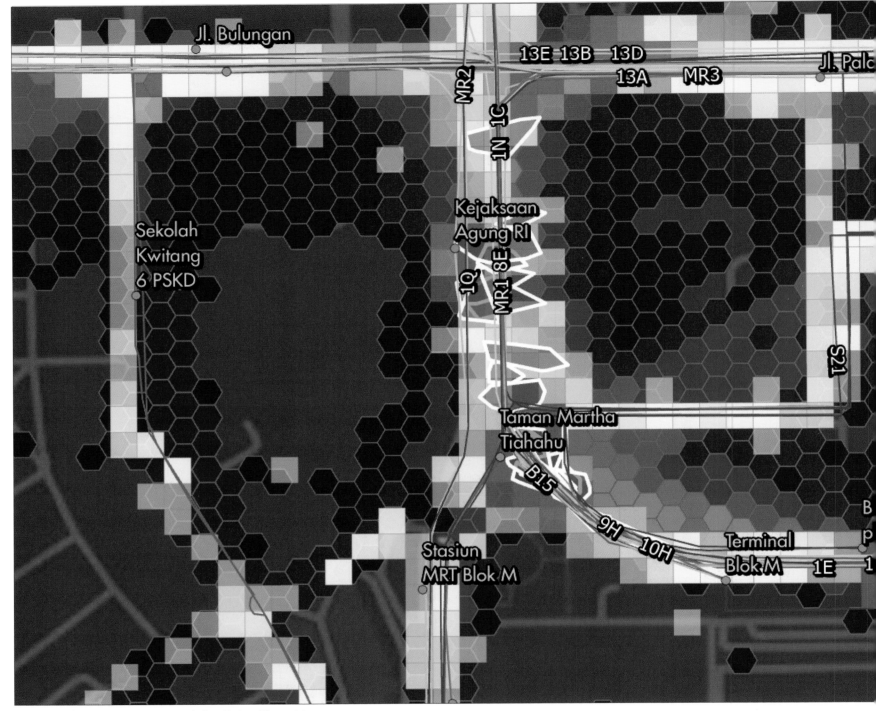

Courtesy of M. Pandito Pratama and Alian Fathira Khomaini.

UNDERSTANDING JAKARTA'S BUS ROUTE CONGESTION

Pt Transportasi Jakarta
Jakarta, Indonesia
By Andika Hadi Hutama

The bus rapid transit (BRT) system in Jakarta is the world's longest system (251.2 km), with 4,000 buses and 1 million passengers daily. Geo-enabling the system helps mitigate issues related to traffic. Analyzing the

GPS records of bus movement revealed bus location patterns during peak commuting times and showed whether congestion was caused by specific incidents.

CONTACT
Andika Hadi Hutama
ahhutama@esriindonesia.co.id

SOFTWARE
ArcGIS Pro, ArcGIS
Online, ArcGIS Desktop
GeoAnalytics

DATA SOURCES
Transjakarta

Courtesy of M. Pandito Pratama and Alian Fathira Khomaini.

SEEING COMMUTER TRAVEL PATTERNS

Pt Transportasi Jakarta
Jakarta, Indonesia
By Andika Hadi Hutama

Commuters in Jakarta use prepaid cards or e-tickets when entering or leaving the bus system. By leveraging this "tap in, tap out" data across the whole network, analysts can conduct origin–destination (OD) modeling to analyze and observe patterns in the flow of people. Putting the OD model on a map provides a straightforward understanding of business operations. This analysis can later be enhanced by superimposing relevant information, such as demographics and points of interest, to understand travel patterns.

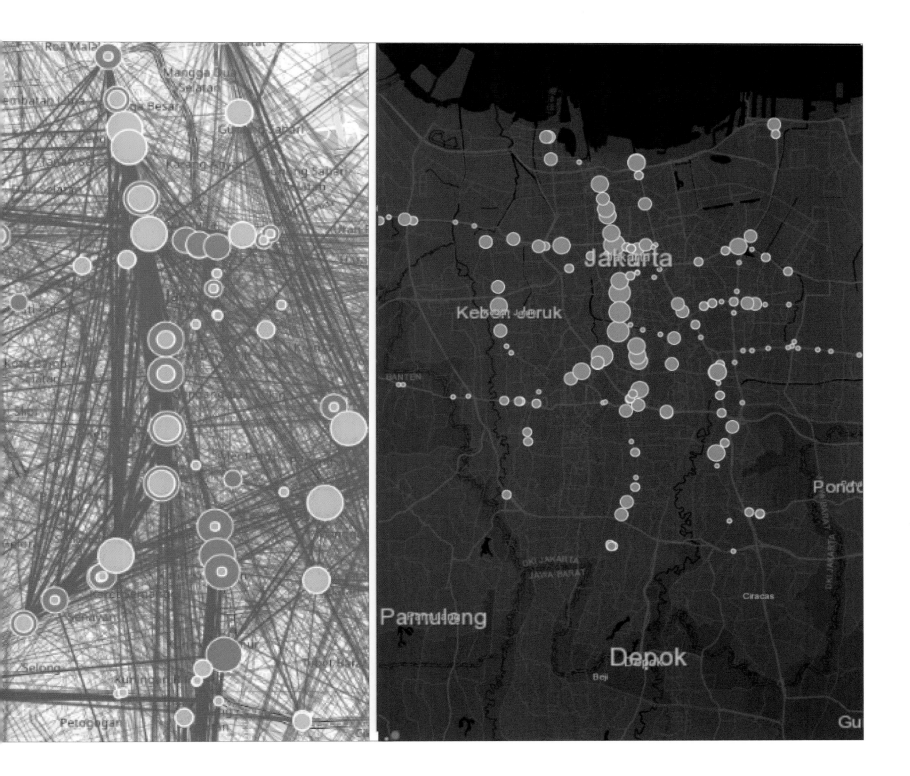

CONTACT
Andika Hadi Hutama
ahhutama@esriindonesia.co.id

SOFTWARE
ArcGIS Pro,
ArcGIS® Insights℠

DATA SOURCES
Transjakarta

ECONOMIC OUTCOMES OF DEVELOPMENT PATTERNS

University of Tennessee, Department of Geography
Knoxville, Tennessee, USA
By Hunter Sinclair

This overhead view of tax value versus land area of parcels shows stakeholders the low tax productivity of low-density sprawl contrasted with the higher tax productivity of high-density, mixed-use development in the urban core. Demonstrating this pattern shows stakeholders that transit-oriented development makes good administrative and financial sense.

CONTACT

Hunter Sinclair
hsinclai@yahoo.com

SOFTWARE

ArcGIS Pro, ArcGIS Urban, ArcGIS 3D Analyst

DATA SOURCES

Knoxville Geographic Information Service, Nearmap 3D Mesh

Courtesy of University of Tennessee, Department of Geography.

E. Jackson Avenue

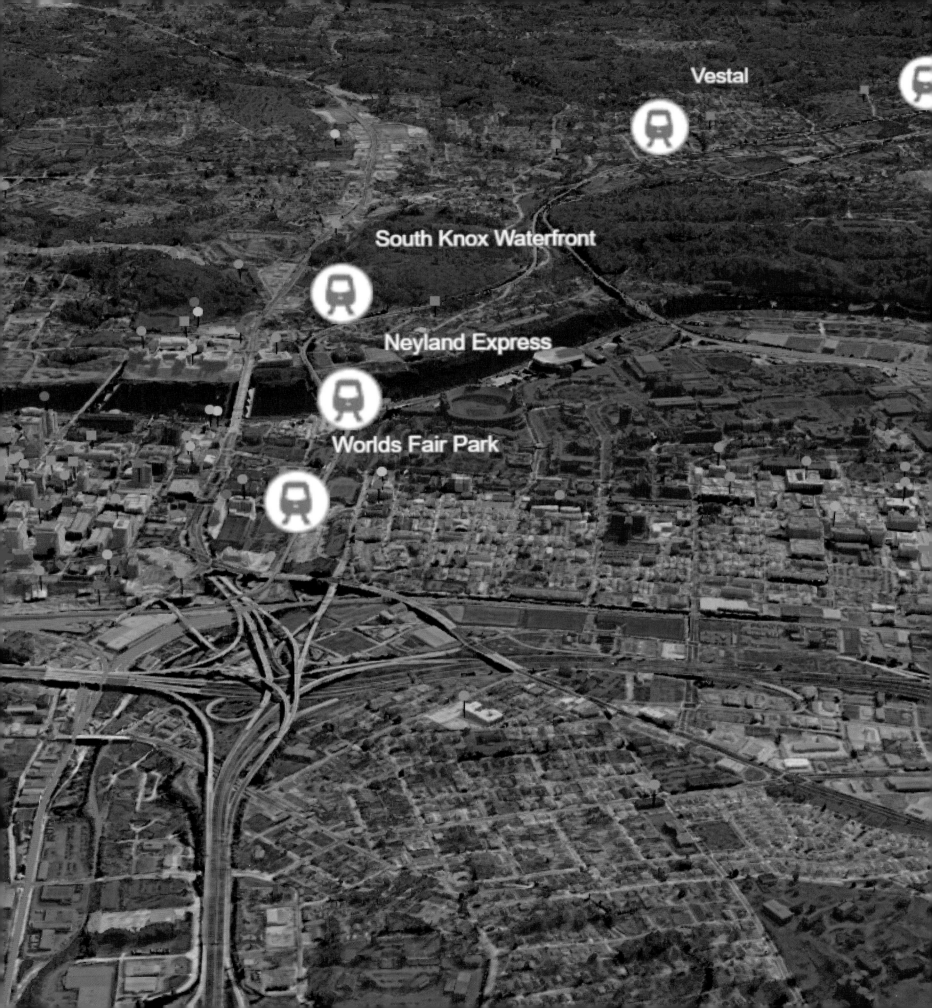

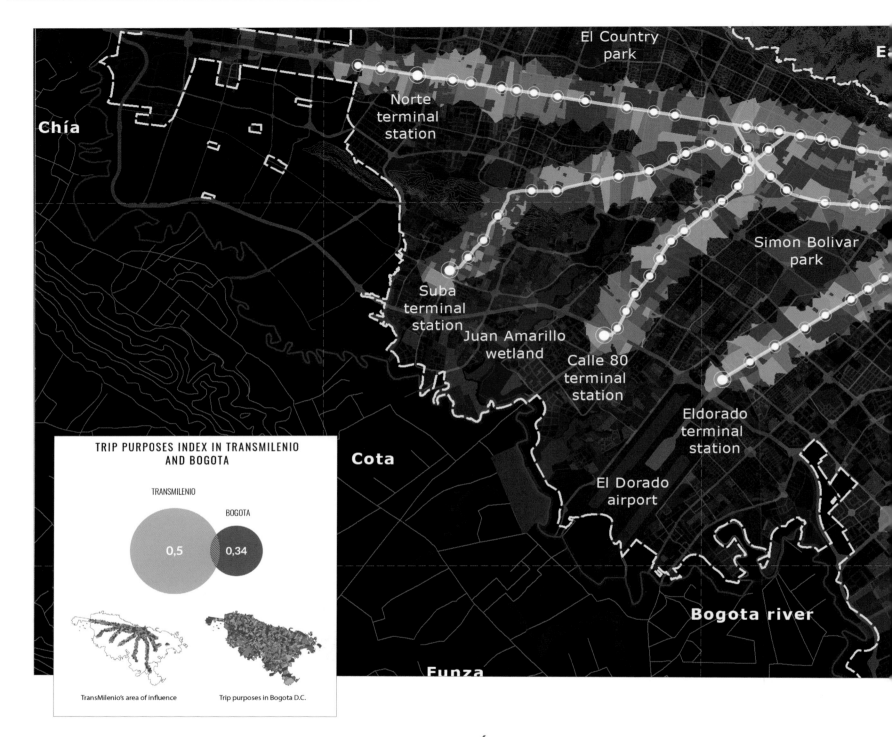

TRIP PURPOSES INDEX IN TRANSMILENIO
AND BOGOTA

TRANSMILENIO

BOGOTA

0,5

0,34

TransMilenio's area of influence

Trip purposes in Bogota D.C.

DIVERSITY OF TRIP PURPOSES IN BOGOTÁ, COLOMBIA

TransMilenio S. A.
Bogotá, Colombia
By Manuel Camilo Chala
Penagos

Every trip has a mode and a purpose, indicating the dynamics of a city. This map shows trips in Bogotá, Colombia, based on biodiversity theory, which calculates the proportion of individuals of any species with respect to the total number of individuals. In this case, the map analyzes the purposes of traveling in each

neighborhood (commuting, studying, leisure, or other) as "the species," and considers the number of trips for each one of them, using the travel demand survey of 2015.

The blue areas are more desirable because there are many reasons to travel there. One could visit a doctor's office, go to work, and then see a movie, without traveling

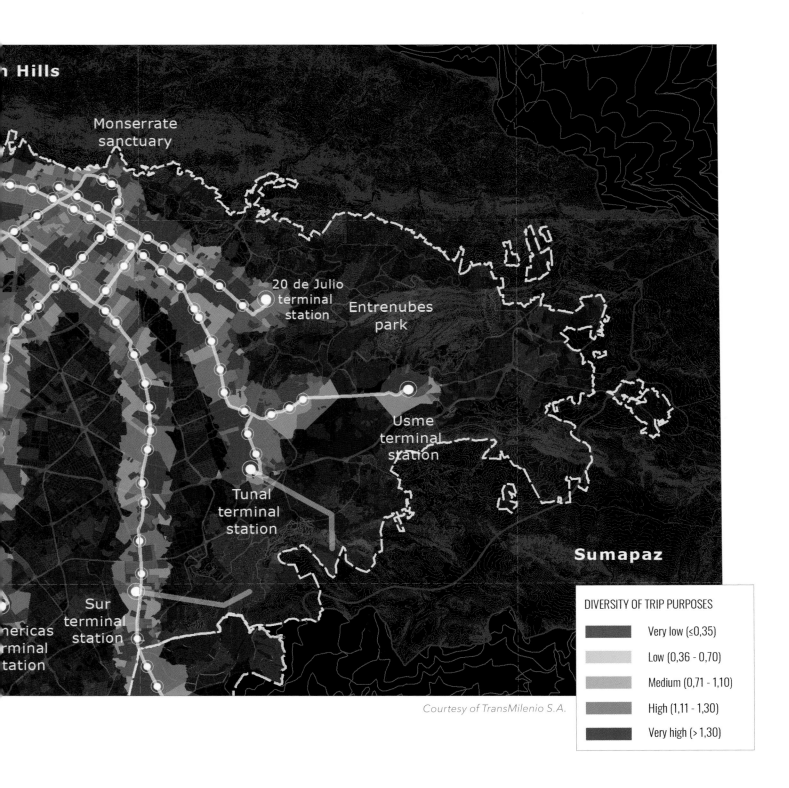

Courtesy of TransMilenio S.A.

DIVERSITY OF TRIP PURPOSES

	Very low (≤0,35)
	Low (0,36 - 0,70)
	Medium (0,71 - 1,10)
	High (1,11 - 1,30)
	Very high (> 1,30)

far. These areas were compared with all areas within 500 meters of the bus rapid transit system of Bogotá. The result showed that the blue areas were needed for trips 1.5 times more than the city average.

CONTACT
Manuel Camilo Chala Penagos
manuel.chala@transmilenio.gov.co

SOFTWARE
ArcGIS Desktop,
ArcGIS Spatial Analyst,
Adobe InDesign

DATA SOURCES
TransMilenio S. A.,
Secretary of Mobility
of Bogotá

AIRPORT 3D CAMPUS AND GATE WALK TIMES WITH CONSTRUCTION IMPACTS

San Francisco International Airport (SFO)
San Francisco, California, USA
By Hanson Guy Michael and Agie Gilmore (SFO 3D campus); Hanson Guy Michael, Agie Gilmore, and Aric Lang (gate walk times)

During the ongoing capital improvement project at San Francisco International Airport, the GIS team received dozens of building information modeling (BIM) datasets to develop a 3D model of the airport, including interiors, exteriors, and underground utilities. This allows users to visualize indoor spaces for the purposes of wayfinding, security operations, maintenance, and planning, all within a web browser.

Analysis was used to estimate travel time to connecting flights during construction of the new Harvey Milk Terminal 1. It considered elevators, escalators, moving walkways, and the SFO AirTrain system to route travelers from every permutation of one gate to another. Calculations considered average wait times at security checkpoints, elevators, average walk speeds, and escalator versus stair speeds.

CONTACT
Hanson Guy Michael
guy.michael@flysfo.com

SOFTWARE
ArcGIS Pro, ArcGIS Enterprise portal (SFO 3D campus); ArcGIS Pro, ArcGIS Network Analyst (gate walk times)

DATA SOURCES
SFO GIS, Nearmap, Esri (SFO 3D campus); SFO GIS (gate walk times)

Courtesy of San Francisco International Airport.

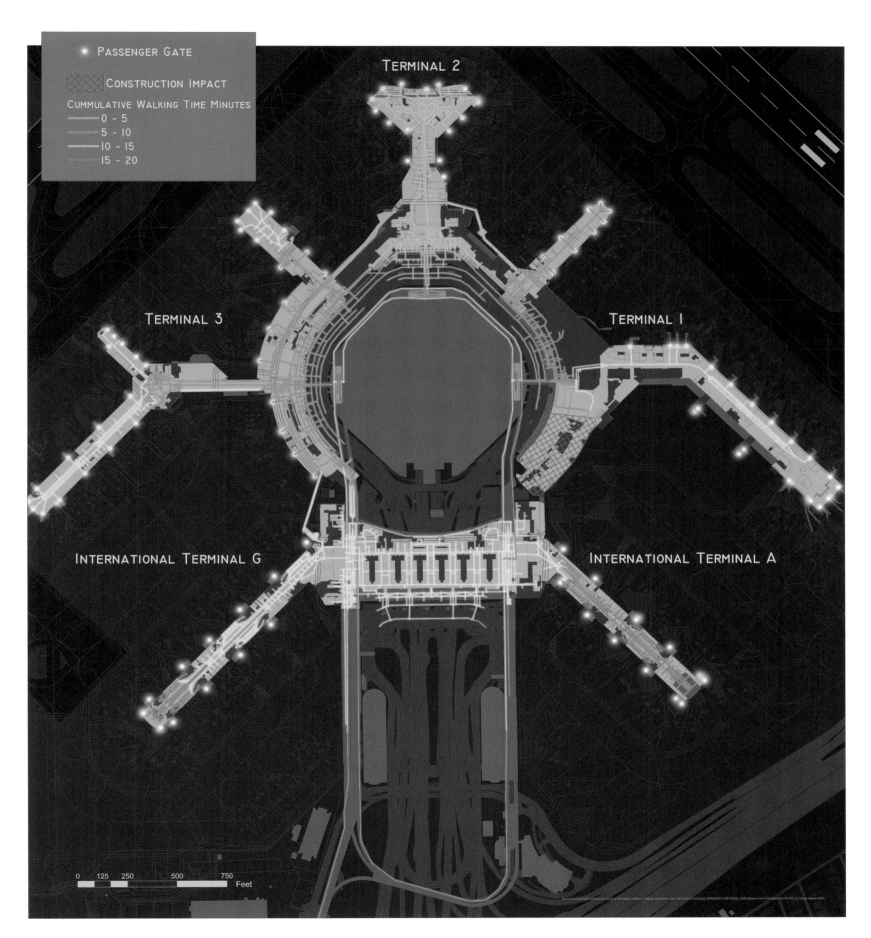

COLONIAL PIPELINE CYBERATTACK OUTAGES

Northeastern Pennsylvania (NEPA) Alliance
Pittson, Pennsylvania, USA
By Annette Ginocchetti

The Colonial Pipeline Gas Buddy Station Outages application highlights the location of the Colonial Pipeline in relation to crowdsourced gas station outages. This layer displays outages due to the Colonial Pipeline ransomware attack that shut down the major gasoline and jet fuel pipeline to large swaths of the South and East Coast, leading to a temporary gas shortage. In the event that the gas shortage made its way into Northeastern Pennsylvania, the NEPA Alliance would be able to make informed decisions and alert the public.

CONTACT
Annette Ginocchetti
aginocchetti@nepa-alliance.org

SOFTWARE
ArcGIS Online, ArcGIS Web AppBuilder

DATA SOURCES
Petroleum Products Pipelines feature layer by Esri federal user community, Gas Buddy Retail Gas Station Crowdsourced Outages feature layer by FEMA HQ GeoEvent

Courtesy of the Northeastern Pennsylvania Alliance.

'DIGITAL WALL MAP' FOR THE PENNSYLVANIA PIPELINE PROJECT

Tetra Tech Inc.
Englewood, Colorado, USA
By Roxanne Burridge and Mike Wager

Tetra Tech developed a powerful web map for a 360-mile linear pipeline project in Pennsylvania. This project had more than 100 complex map layer sets, thousands of personnel, and more than 25,000 documents and drawings that needed to be accessed for the full life cycle of the project, from bid to as built. In the early stages of the project, data sharing was handled by sending large KMZ files, which quickly became out of date, leading to confusion and elevating the risk of potential errors. This led to a vision to create a "digital wall map."

The web map had hyperlinks to the necessary drawings and documents, providing the ability to monitor construction progress without the added cost and time of printing. Changes were reflected in real time to all users, updating as pipes were laid. The web map hosted 2 cm orthoimagery and 1 ft. lidar contours. Additionally, the web map met rigorous environmental and regulatory documentation requirements by providing vital information on wetlands, streams, and environmentally sensitive areas.

CONTACT
Roxanne Burridge
roxanne.burridge@tetratech.com

SOFTWARE
ArcGIS Desktop, ArcGIS Pro, ArcGIS Web AppBuilder, ArcGIS Enterprise portal

DATA SOURCES
Wetlands, streams, orthoimagery, and lidar collected by Tetra Tech

Courtesy of Tetra Tech Inc.

INDICATORS OF BROADBAND NEED IN THE US

National Telecommunications and Information Administration (NTIA), US Department of Commerce
Washington, DC, USA
By Lindsey Hays

This Indicators of Broadband Need map was created by the Department of Commerce, NTIA. The map uses several data sources to show information on broadband availability within the United States.

Layers in this map were created using data sourced from the American Community Survey collected by the US Census, Measurement Lab (M-Lab), Ookla, Microsoft, and the Federal Communications Commission (FCC). In addition, layers in the map display the locations of higher education institutions eligible as minority-serving institutions (MSIs) and show areas designated as American Indian, Alaska Native, and Native Hawaiian Areas by the US Census in 2020.

CONTACT
Lindsey Hays
lhays@ntia.gov

SOFTWARE
ArcGIS Online

DATA SOURCES
Census American Community Survery, FCC Form 477, Measurement Lab, Ookla Speedtest, Microsoft

Courtesy of National Telecommunications and Information Administration and US Department of Commerce.

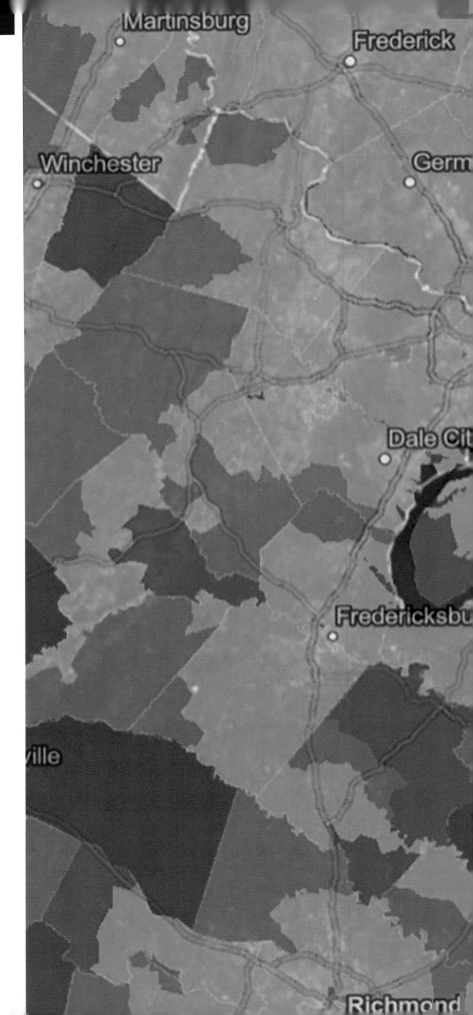

LOCATING UNDERGROUND INFRASTRUCTURE

CyberTech Systems and Software Inc.
Oak Brook, Illinois, USA
By Sagar Nair

Spatialitics Line Locate is a modern solution for locating buried utility assets such as water and gas lines. The solution allows the public to request line location services, efficiently allocates field personnel to service locations, and reduces locate time from days to minutes.

This map highlights part of the underground infrastructure for a major water company and shows the locations where the infrastructure needs to be indicated on the ground. Field personnel use the app to get an optimized route for their service calls and mark them complete.

CONTACT

Sagar Nair
sagar.nair@cybertech.com

SOFTWARE

ArcGIS API for JavaScript, ArcGIS Desktop, ArcGIS Pro

DATA SOURCES

CyberTech Systems

Courtesy of CyberTech Systems and Software Inc.

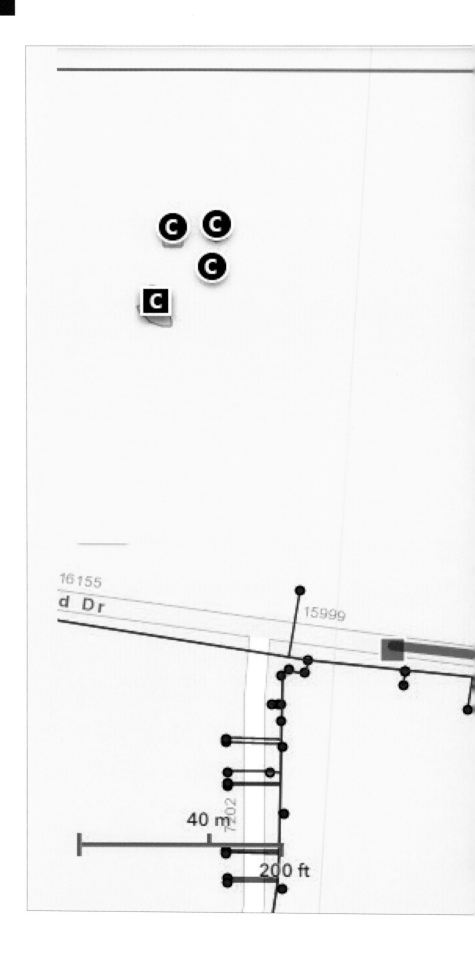

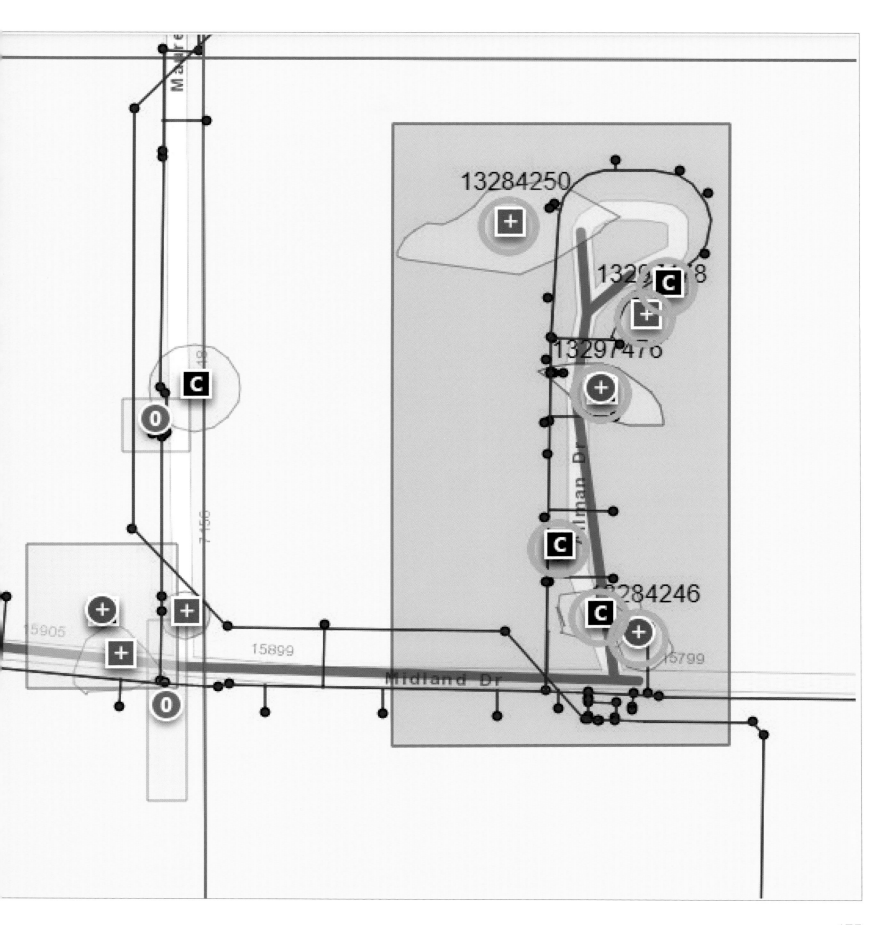

KEEPING TRAFFIC FLOWING IN THE CITY OF CARLSBAD

City of Carlsbad
Carlsbad, California, USA
By David Young

The communication system for the City of Carlsbad's traffic network was managed by an outside consultant for many years, but as the city's Information Technology (IT) Department began to implement Carlsbad's fiber-based Digital Information Network (CDIN), it became apparent that IT should assume the primary responsibility for managing it.

IT approached the city's GIS team about creating an interactive web map that would display the communication network and serve as a collaboration tool. Visualizing the communication network on a map quickly revealed patterns and choke points that would have otherwise been difficult to detect and allowed IT to address those problem areas. In addition, the ability to visualize wired versus wireless network elements enhanced remediation efforts for the existing network and improved the planning process for future network upgrades. IT now has an easy-to-use tool for visualizing and analyzing the city's traffic network infrastructure.

CONTACT
David Young
david.young@carlsbadca.gov

SOFTWARE
ArcGIS Enterprise portal, ArcGIS Desktop

DATA SOURCES
City of Carlsbad

Courtesy of City of Carlsbad.

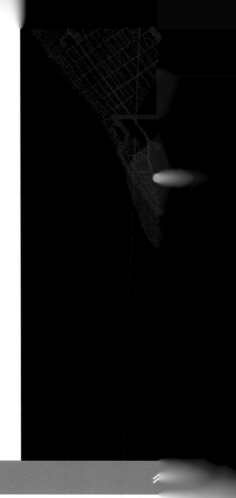

Legend

Traffic Network - Signal Cabinet (Contractor Data)

- ● Group 1
- ● Group 2
- ● Group 3
- ● Group 4
- ● Group 5
- ● Group 6
- ● Group 7
- ● Group 8
- ● Group 9
- ● Group 10
- ● Group 11
- ⊗ CalTrans
- ○ Future
- ◉ Off Line

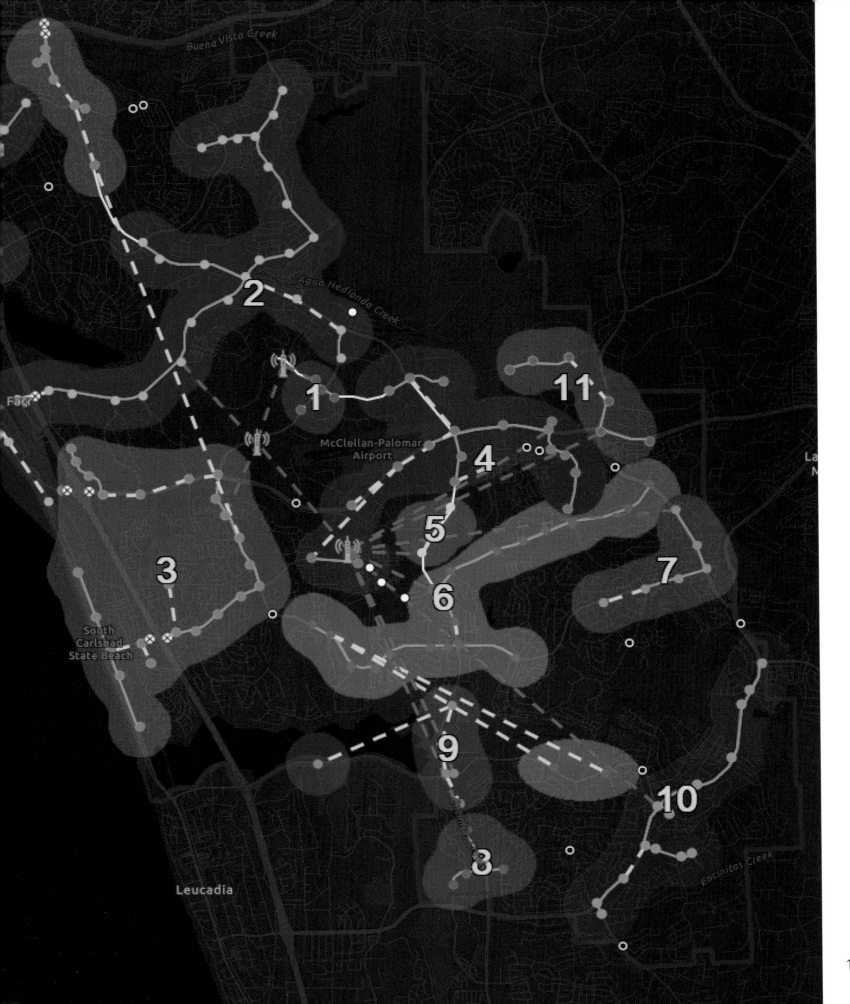

STREETLIGHT MAINTENANCE IN DURHAM, NORTH CAROLINA

City of Durham
Durham, North Carolina, USA
By Robert Cushman

Street lighting is a shared responsibility in Durham, North Carolina. Some streetlights are maintained by private owners, some by the North Carolina Department of Transportation, and the balance by the City of Durham.

City staff use this single-focus map to coordinate with both Duke Energy and Durham residents to quickly repair or replace light fixtures when needed. The map lets staff know who is responsible for the maintenance of each light and reveals additional details about each light fixture. Database security features keep this powerful resource secure and available 24/7.

CONTACT
Robert Cushman
rob.cushman@durhamnc.gov

SOFTWARE
ArcGIS Pro, ArcGIS Enterprise

DATA SOURCES
City of Durham, Duke Energy Corporation

Courtesy of City of Durham.

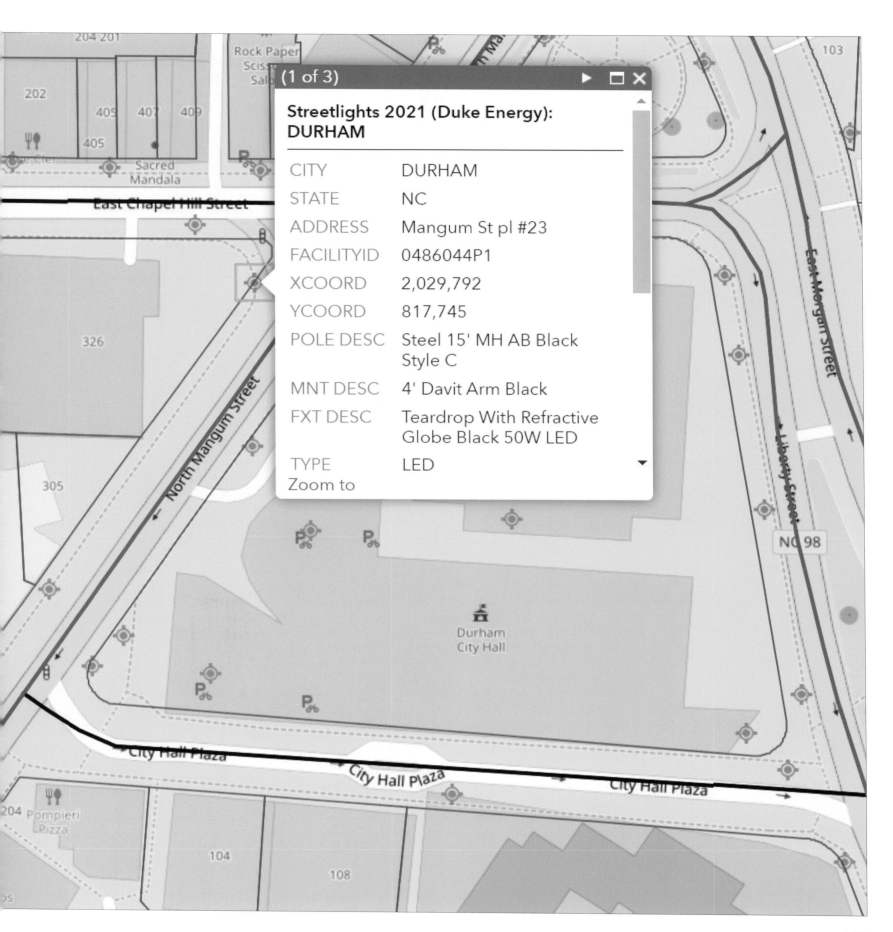

F2500
WA160-21

F2410
T1-27

F2400
WA160-21

Courtesy of K2 Engineering GmbH.

VISUALIZING A PLANNED POWER LINE

K2 Engineering GmbH
Leipzig, Germany
By Jens Jacob

This map compares two proposals for a planned power line, one through a forest and the other with a cutting lane.

F2410
T1-45

F2510
T1-43

F2400
WA160-21

CONTACT
Jens Jacob
Jens.Jacob@K2E.de

SOFTWARE
ArcGIS Data Interoperability,
ArcGIS Online, ArcGIS Pro,
VISALL, Microstation, Civil 3D

DATA SOURCES
OpenStreetMap, Esri World
Imagery, Geobasisdaten, Bayerische
Vermessungsverwaltung, CityGML

SOLAR PANELS NEXT TO ROADS

The Ray, Atlanta, Georgia, USA
By Laura Rogers

This right-of-way (ROW) solar tool was developed to analyze and plan solar development on highway roadsides. It allows the Iowa Department of Transportation to estimate energy generation, economic value, and greenhouse gas equivalencies.

Courtesy of The Ray.

CONTACT
Laura Rogers
laura@theray.org

SOFTWARE
ArcGIS Pro

DATA SOURCES
Iowa Department of Transportation on ArcGIS Online, Iowa State
University GIS Facility on ArcGIS Online, ArcGIS Online, Homeland
Infrastructure Foundation-Level Data, Environmental Protection Agency,
GeoTREE, University of Texas at Austin, USDA, USGS

MAPPING THE INCREASE OF ELECTRIC VEHICLES IN HAWAII

4CTechnologies, Verona, Pennsylvania, USA
By Susan Zwillinger

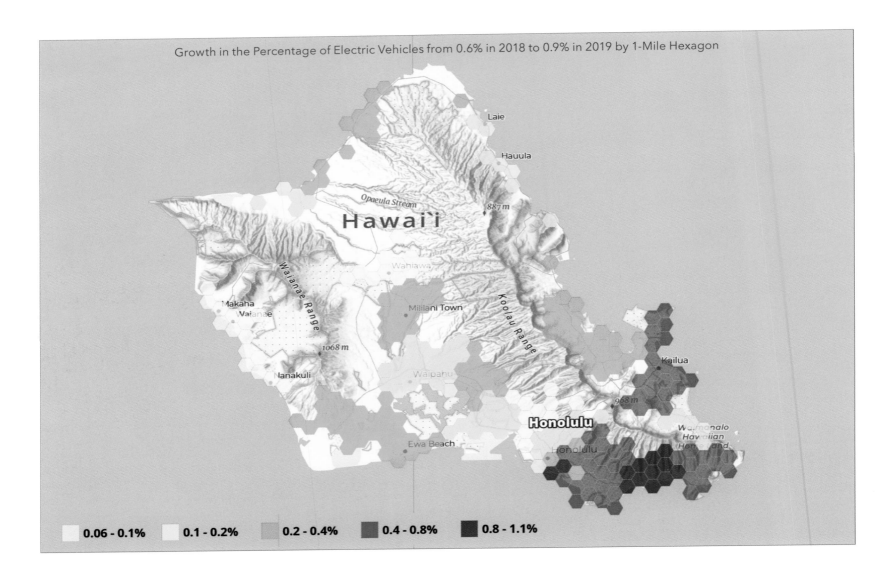

Growth in the Percentage of Electric Vehicles from 0.6% in 2018 to 0.9% in 2019 by 1-Mile Hexagon

0.06 - 0.1%   0.1 - 0.2%   0.2 - 0.4%   0.4 - 0.8%   0.8 - 1.1%

Electric vehicles are part of a sustainable future, and Hawaii has seen significant growth in electric vehicles (64%) from 2018 to 2019. This study shows how adoption rates vary within the state.

Mapping the total number of electric vehicles shows a pattern that is closely correlated with population density. The city of Honolulu has the highest number of electric vehicles in the state, but the percentage of electric vehicles varies within the city. In some areas, the overall number of registered vehicles declined but the number of electric vehicles increased.

The uplift pattern is easily recognized when a 1-mile hexagon is used. Vehicle registrations by zip code were apportioned to the hexagons using the census block apportionment method. To understand the marketing uplift, the maps show the number of vehicles added from the previous year and the percentage change. The overall percentage of electric vehicles for the island of Hawaii is still low—not even a full 1% of vehicles are electric, but the growth from 2018 to 2019 shows that adoption is improving.

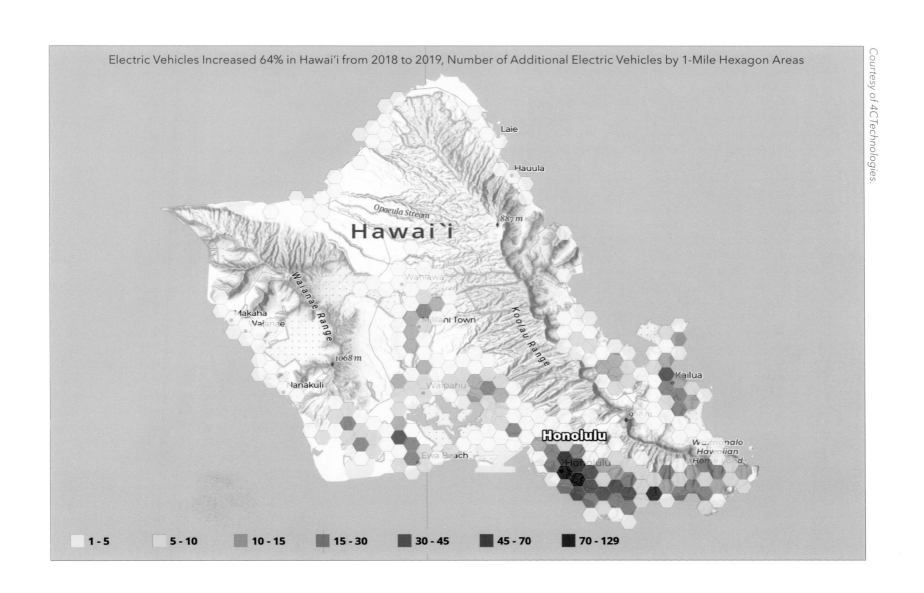

Electric Vehicles Increased 64% in Hawai'i from 2018 to 2019, Number of Additional Electric Vehicles by 1-Mile Hexagon Areas

Courtesy of 4CTechnologies.

Legend: 1 - 5 | 5 - 10 | 10 - 15 | 15 - 30 | 30 - 45 | 45 - 70 | 70 - 129

CONTACT
Susan Zwillinger
szwillinger@4cgeoworks.com

SOFTWARE
ArcGIS Pro,
Esri Demographics,
ArcGIS Business Analyst Desktop

DATA SOURCES
Vehicle registration data from
IHS Polk by ZIP Code, Q1 2020

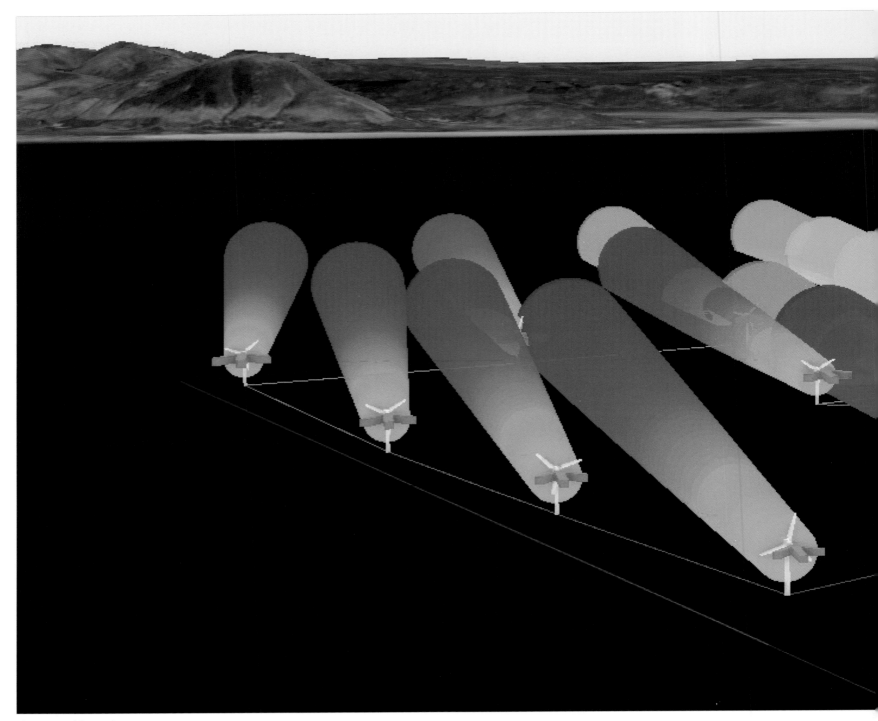

Courtesy of Exprodat.

OPTIMIZING OFFSHORE WIND FARM DESIGN

Exprodat
Irish Sea, United Kingdom
By Adrian Birch, Chris Jepps,
and Richard Webb

Finding the most efficient layout for wind turbines and related network infrastructure for offshore wind installations can be a challenge. Wake effects, caused when one turbine interferes with the available wind for another turbine, can reduce efficiency, so modeling their placement is key.

This scene shows wind turbine locations and 3D-modeled wake effects, highlighting potential wind speed loss for an offshore installation awarded to BP and Energie Baden-Wuerttemberg during the 4th UK Offshore Wind Round, in February 2021.

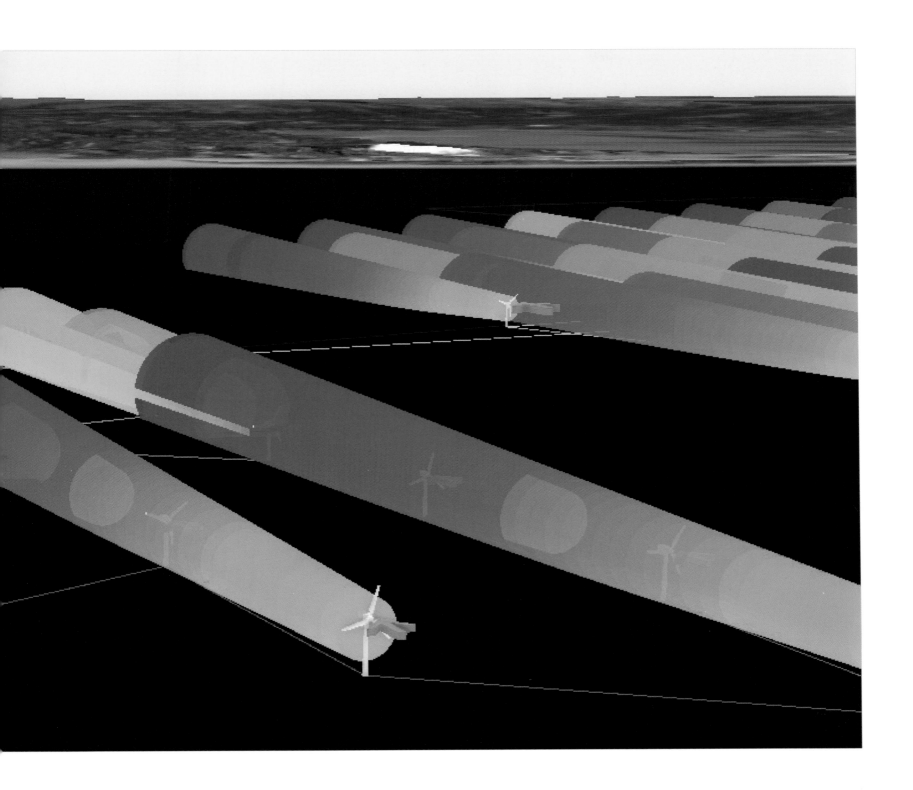

CONTACT
Exprodat
info@exprodat.com

SOFTWARE
ArcGIS Pro, ArcGIS Pro SDK

DATA SOURCES
Esri World Imagery

ASSESSING 5G COVERAGE IN URBAN ENVIRONMENTS

UAB Ekspera LT, Vilnius, Lithuania
By Vytautas Ramonaitis

5G brings many benefits, including higher capacity and throughput with lower latency. Beamforming, directing a signal at a specific device, is one of the leading technologies that enables those benefits. However, it is challenging to model the signal propagation of beamforming antennas, especially in urban environments.

This image shows 5G coverage in downtown Orlando, Florida. Lidar data was used for signal propagation analysis, overlaid on ArcGIS Online basemaps and 3D building data. This synchronized side-by-side view indicates 2D signal coverage at the ground level, left, and 3D signal propagation, right.

CONTACT
Vytautas Ramonaitis
vramonaitis@expera.online

SOFTWARE
ArcGIS Pro, ArcGIS Spatial
Analyst, ArcGIS Online

DATA SOURCES
ArcGIS Online

GAS UTILITY NETWORK IN LITHUANIA

Energijos Skirstymo Operatorius (ESO)
Klaipėda, Lithuania
By Žygintas Žalys

This map shows a snapshot of the gas utility network in Lithuania. It is used daily by many workers who view, analyze, or update the data. An offline map is also available to collect data in regions where there is no internet coverage.

CONTACT
Žygintas Žalys
zygintas.zalys@eso.lt

SOFTWARE
ArcGIS Pro, ArcGIS Utility Network

DATA SOURCES
Gas utility network service

Courtesy of ESO.

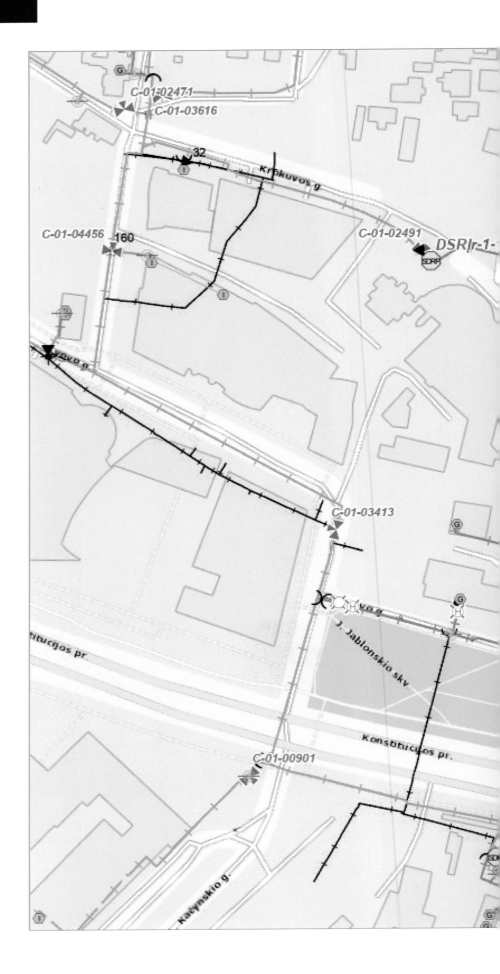

C-01-02523
Krokuvos g.
Krokuvos g.
Kalvarijų g.
Vilniaus vaikų
lopšelis-darželis
PIPIRAS
S-01-00651
DSR|r-1-49
S-01-03495
S-01-04545
Vilniaus vaikų
lopšelis-darželis
DRUGELIS
C-01-03355
Kalvarijų g.
Piromontas
Šeimyniškių g.
Šeimyniškių g.
Šeimyniškių g.
Slucko g.
1-161

INDEX BY ORGANIZATION